Applied Colloid and Surface Chemistry

Applied Colloid and Surface Chemistry

Second Edition

RICHARD M. PASHLEY
University of New South Wales Canberra
Australia

MARILYN E. KARAMAN
University of New South Wales Canberra
Australia

WILEY

Registered Offices
John Wiley & Sons, Inc., 111 River Street, Hoboken, NJ 07030, USA
John Wiley & Sons Ltd, The Atrium, Southern Gate, Chichester, West Sussex, PO19 8SQ, UK

Editorial Office
The Atrium, Southern Gate, Chichester, West Sussex, PO19 8SQ, UK

For details of our global editorial offices, customer services, and more information about Wiley products visit us at www.wiley.com.

Wiley also publishes its books in a variety of electronic formats and by print-on-demand. Some content that appears in standard print versions of this book may not be available in other formats.

Library of Congress Cataloging-in-Publication Data

Names: Pashley, Richard M., author. | Karaman, Marilyn E., author.
Title: Applied colloid and surface chemistry / Richard M. Pashley,
    University of New South Wales Canberra, Canberra, Australia, Marilyn E.
    Karaman, University of New South Wales Canberra, Canberra, Australia.
Description: Second edition. | Hoboken, NJ : Wiley, 2021. | Includes
    bibliographical references and index.
Identifiers: LCCN 2021031901 (print) | LCCN 2021031902 (ebook) | ISBN
    9781119739128 (paperback) | ISBN 9781119740001 (adobe pdf) | ISBN
    9781119740018 (epub)
Subjects: LCSH: Colloids. | Surface chemistry.
Classification: LCC QD549 .P275 2021 (print) | LCC QD549 (ebook) | DDC
    541/.345–dc23
LC record available at https://lccn.loc.gov/2021031901
LC ebook record available at https://lccn.loc.gov/2021031902

Cover Design: Wiley
Cover Image: © Photo-Max/E+/Getty Images

Set in 10/13pt Sabon LT Std by Straive, Pondicherry, India

Those that can, teach.

Sit down before fact as a little child, be prepared
to give up every preconceived notion, follow
humbly wherever and to whatever abysses nature
leads, or you shall learn nothing.

Thomas Henry Huxley (1860)

Contents

Preface

The first edition of this book was written following several years of teaching this material to third year undergraduate and honours students in the Department of Chemistry at the Australian National University in Canberra, Australia. Science students are increasingly interested in the application of their studies to the real world, and colloid and surface chemistry is an area which offers many opportunities to apply learned understanding to everyday and industrial examples. There is a lack of resource materials with this focus, and so we produced the first edition with many industrial examples.

This second edition extends this to include several more recent and topical industrial innovations. It is still intended to take chemistry or physics students with no background in the area to the level where they are able to understand many natural phenomena and industrial processes and are able to consider potential areas of new research. It involves the study of the interaction between solids, liquids and gases and their application to numerous everyday processes.

Colloid and surface chemistry spans the very practical to the very theoretical, and less mathematical students may wish to skip some of the more involved derivations. However, they should be able to do this and still maintain a good basic understanding of the fundamental principles involved. It should be remembered that a thorough knowledge of theory can act as a barrier to progress, through the inhibition of further investigation. Students asking ignorant but intelligent questions can often stimulate valuable new research areas.

The book contains some recommended experiments, which we have found to work well and stimulate students to consider both the fundamental theory and industrial applications. Sample questions have also been included in some sections, with detailed answers available on our web site.

Although the text has been primarily aimed at students, researchers in cognate areas may also find some of the topics stimulating. A reasonable background in chemistry or physics is all that is required.

We also would like to gratefully acknowledge important contributions from several students, including John Antony and Mojtaba Taseidifar (for Chapter 3) and Mathew Francis and Rui Wei (for Chapter 11).

Richard M. Pashley
Marilyn E. Karaman
October 2020

About the Companion Website

This book is accompanied by a companion website.

www.wiley.com/go/pashley/appliedcolloid2e

This website includes:

* Powerpoint slides of Figures from the book

1

Introduction

Introduction to the nature of colloids and the linkage between colloids and surface properties. The importance of size and surface area. Introduction to wetting and the industrial importance of particle size and surface modifications.

INTRODUCTION TO THE NATURE OF COLLOIDAL SOLUTIONS

The difference between macroscopic and microscopic objects is clear from everyday experience. For example, a glass marble will sink rapidly in water; however, if we grind it into sub-micron-sized particles, these will float or disperse freely in water, producing a visibly cloudy 'solution', which can remain stable for hours or days. In this process we have, in fact, produced a 'colloidal' dispersion or solution. This dispersion of one (finely divided or microscopic) phase in another is quite different to the molecular mixtures or 'true' solutions formed when we dissolve ethanol or common salt in water. Microscopic particles of one phase dispersed in another are generally called colloidal solutions or dispersions. Both nature and industry have found many uses for this type of solution. We will see later that the properties of colloidal solutions are intimately linked to the high surface area of the dispersed phase, as well as to the chemical nature of the particle's surface.

Historical note: The term 'colloid' is derived from the Greek word for glue, 'kolla'. It was originally used for gelatinous polymer colloids,

Applied Colloid and Surface Chemistry, Second Edition. Richard M. Pashley and Marilyn E. Karaman.
© 2021 John Wiley & Sons Ltd. Published 2021 by John Wiley & Sons Ltd.
Companion website: www.wiley.com/go/pashley/appliedcolloid2e

which were identified by Thomas Graham in 1860 in experiments on osmosis and diffusion.

It turns out to be very useful to dissolve (or more strictly disperse) solids, such as minerals and metals, in water. But how does it happen? We can see why from simple physics. Three fundamental forces operate on fine particles in solution:

(1) A gravitational force, tending to settle or raise particles depending on their density relative to the solvent.
(2) A viscous drag force, which arises as a resistance to motion, since the fluid has to be forced apart as the particle moves through it.
(3) The 'natural' kinetic energy of particles and molecules, which causes random Brownian motion.

If we consider the first two non-random forces, we can easily calculate the terminal or limiting velocity, V, (for settling or rising, depending on the particle's density relative to, say, water) of a spherical particle of radius r. Under these conditions, the viscous drag force must equal the gravitational force. Thus, at a settling velocity, V, the viscous drag force is given by: $F_{drag} = 6\pi r V \eta = 4\pi r^3 g(\rho_p - \rho_w)/3 = F_{gravity}$, the gravitational force, where η is the viscosity of water and the density difference between particle and water is $(\rho_p - \rho_w)$. Hence, if we assume a particle-water density difference of +1 g cm^{-3}, we obtain the results:

$r/\text{Å}$	100	1000	10,000	10^5	10^6
$r/\mu\text{m}$	0.01	0.1	1	10	100
$V/\text{cm/sec}$	2×10^{-8}	2×10^{-6}	2×10^{-4}	2×10^{-2}	2

Clearly, from factors (1) and (2), small particles will take a very long time to settle, and so a fine dispersion will be stable almost indefinitely, even for materials denser than water. But what of factor (3)? Each particle, independent of size, will have a kinetic energy, on average, of around 1 kT. So the typical random speed (v) of a particle (in any direction) will be roughly given by:

$$mv^2 / 2 \cong 1kT \cong 4 \times 10^{-21} \, \text{J} \, (\text{at room temperature})$$

Again, if we assume that ρ_p = 2 g cm^{-3}, then we obtain the results:

r/Å	100	1000	10,000	10^5	10^6
r/µm	0.01	0.1	1	10	100
v/cm/sec	10^2	3	0.1	3×10^{-3}	1×10^{-4}

These values suggest that kinetic random motion will dominate the behaviour of small particles, which will not settle, and the dispersion will be completely stable. However, this point is really the beginning of 'colloid science'. Since these small particles have this kinetic energy they will, of course, collide with other particles in the dispersion, with collision energies ranging up to at least 10 kT (since there will actually be a distribution of kinetic energies). If there are attractive forces between the particles – as is reasonable since most colloids were initially formed via a vigorous mechanical process of disruption of a macroscopic or large body – each collision might cause the growth of larger aggregates, which will then, for the reasons already given, settle out and we will no longer have a stable dispersion! The colloidal solution will coagulate and produce a solid precipitate at the bottom of a clear solution.

There is, in fact, a ubiquitous force in nature, called the van der Waals (vdw) force, which is one of the main forces acting between molecules and is responsible for holding together many condensed phases, such as solid and liquid hydrocarbons and polymers. It is responsible for about one third of the attractive force holding liquid water molecules together. This force was actually first observed as a correction to the ideal gas equation and is attractive even between neutral gas molecules, such as oxygen and nitrogen, in a vacuum, which is why they can be liquified. Although electromagnetic in origin (as we will see later), it is much weaker than the coulombic force acting between ions, such as in salt crystals.

THE FORCES INVOLVED IN COLLOIDAL STABILITY

Although van der Waals forces will always act to coagulate dispersed colloids, it is possible to generate an opposing repulsive force of comparable strength. This force arises because most materials, when dispersed in water, ionise to some degree or selectively adsorb ions from the solution and hence become charged. Two similarly charged colloids will repel each other via an electrostatic repulsion, which will oppose coagulation. The stability of a colloidal solution is therefore critically dependent on the charge generated at the surface of the particles. The combination of these two forces, attractive van der Waals and repulsive electrostatic forces, forms the fundamental basis for our understanding of the behaviour and stability of colloidal solutions. The corresponding theory is referred to as the DLVO (after Derjaguin, Landau, Verwey and Overbeek) theory of colloid stability, which we will consider in greater detail later. The stability of any colloidal dispersion is thus determined by the behaviour of the surface of the particle via its surface charge and its short-range attractive van der Waals force.

Our understanding of these forces has led to our ability to selectively control the electrostatic repulsion and so create a powerful mechanism for controlling the properties of colloidal solutions. As an example, if we have a valuable mineral imbedded in a quartz rock, grinding the rock will separate out both pure individual quartz and the mineral particles, which can both be dispersed in water. The valuable mineral can then be selectively coagulated, whilst leaving the unwanted quartz in solution. This process is used widely in the mining industry as the first stage of mineral separation. The alternative of chemical processing, for example, by dissolving the quartz in hydrofluoric acid, would be both expensive and environmentally unfriendly.

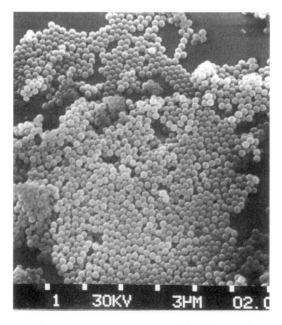

Figure 1.1   Scanning electron microscope image of dried mono-disperse silica colloids.

It should be realised, at the outset, that colloidal solutions (unlike true solutions) will almost always be in a metastable state. That is, an electrostatic repulsion prevents the particles from combining into their most thermodynamically stable state of aggregation into the macroscopic form, from which the colloidal dispersion was (artificially) created in the first place. On drying, colloidal particles will often remain separated by these repulsive forces, as illustrated by the scanning electron microscope picture of mono-disperse silica colloids.

TYPES OF COLLOIDAL SYSTEMS

The term 'colloid' usually refers to particles within the approximate size range of 50 Å to 50 μm, but this, of course, is somewhat arbitrary. For example, blood could be considered as a colloidal solution in which large blood cells are dispersed in water. They are stabilized by their negative charge, and so oppositely charged ions, for example, those produced by an alum stick, will coagulate the cells and hence stop bleeding. Often we are interested in solid dispersions in aqueous solutions, but many other situations are also of interest and industrial importance. Some examples are given in the following table.

The properties of colloidal dispersions are intimately linked to the high surface area of the dispersed phase and the chemistry of these interfaces. This linkage is well illustrated by the titles of two of the main journals in this

Table 1.1

Dispersed phase	Dispersion medium	Name	Examples
Liquid	Gas	Liquid aerosol	Fogs, sprays
Solid	Gas	Solid aerosol	Smoke, dust
Gas	Liquid	Foam	Foams
Liquid	Liquid	Emulsion	Milk, mayonnaise
Solid	Liquid	'Sol', Paste at high concentration	Au sol, AgI sol. Toothpaste
Gas	Solid	Solid foam	Expanded polystyrene
Liquid	Solid	Solid emulsion	Opal, pearl
Solid	Solid	Solid suspension	Pigmented plastics

area: the *Journal of Colloid and Interface Science* and *Colloids and Surfaces*. The natural combination of colloid and surface chemistry represents a major area of both research activity and industrial development. It has been estimated that something like 20 per cent of all chemists in industry work in this area. The more recent term nano is also applied to these small scale materials, because of their typical nanometre size.

THE LINK BETWEEN COLLOIDS AND SURFACES

The link between colloids and surfaces follows naturally from the fact that particulate matter has a high surface area to mass ratio. The surface area of a 1 cm diameter sphere ($4\pi r^2$) is 3.14 cm^2, whereas the surface area of the same amount of material but in the form of 0.1 micron diameter spheres (i.e., the size of the particles in latex paint) is 314,000 cm^2. The enormous difference in surface area is one of the reasons why the properties of the surface become very important for colloidal solutions. One everyday example is that organic dye molecules or pollutants can be effectively removed from water by adsorption onto particulate or granular activated charcoal because of its high surface area. This process is widely used for water purification and in the oral treatment of poison victims.

Although it is easy to see that surface properties will determine the stability of colloidal dispersions, it is not so obvious why this can also be the case for some properties of macroscopic objects. As one important illustration, consider the interface between a liquid and its vapour:

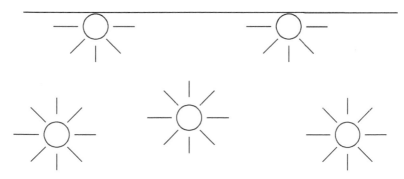

Figure 1.2 Schematic diagram to illustrate the complete bonding of liquid molecules in the bulk phase but not at the surface.

Table 1.2

Liquid	Surface Energy in mJm^{-2} (at 20 °C)	Type of Intermolecular Bonding
Mercury	485	metallic
Water	72.8	hydrogen bonding + vdw
n-Octanol	27.5	hydrogen bonding + vdw
n-Hexane	18.4	vdw
Perfluoro-octane	12	weak vdw

Molecules in the bulk of the liquid can interact via attractive forces (e.g., van der Waals) with a larger number of nearest neighbours than those at the surface. The molecules at the surface must therefore have a higher energy than those in bulk, since they are partially freed from bonding with neighbouring molecules. Thus, work must be done to take fully interacting molecules from the bulk of the liquid to create a new surface. This work gives rise to the surface energy or tension of a liquid. Hence, the stronger the intermolecular forces between the liquid molecules, the greater this work will be, as is illustrated in the table.

The influence of this surface energy can also be clearly seen on the macroscopic shape of liquid droplets, which in the absence of all other forces will always form a shape of minimum surface area – that is, a sphere in a gravity-free system. This is the reason why small mercury droplets are always spherical. Note that the term 'interface' is often used where the surface is formed between two different materials.

Although a liquid will always try to form a minimum surface area shape, if no other forces are involved, it can also interact with other macroscopic objects to reduce its surface tension via molecular bonding to another material, such as a suitable solid. Indeed, it may be energetically favorable for the liquid to interact and 'wet' another material. The wetting properties of a liquid on a particular solid are very important in many everyday activities and are determined solely by surface properties, which are derived from intermolecular forces. One important and common example is that of the behaviour of water on clean glass. Water wets clean glass because of the favourable hydrogen bond interaction between the surface silanol groups on glass and adjacent water molecules, as illustrated below.

However, exposure of glass to Me$_3$SiCl vapour rapidly produces a 0.5 nm layer of methyl groups on the surface:

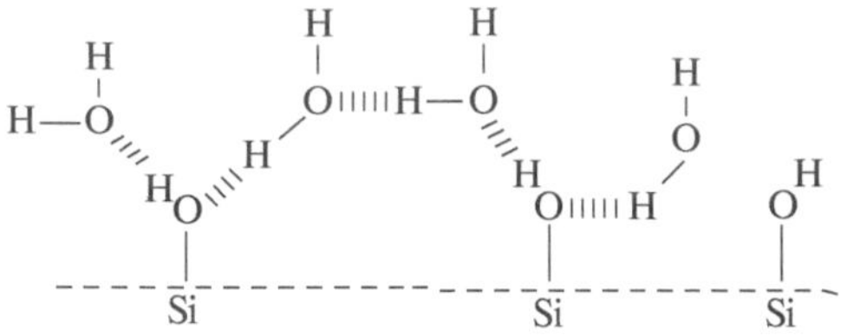

H—O⟍ ... O⟋ ⁗H—O⟍ ... H O H
(Figure 1.3 structure)

Figure 1.3 Water molecules form hydrogen bonds with the silanol groups at the surface of clean glass.

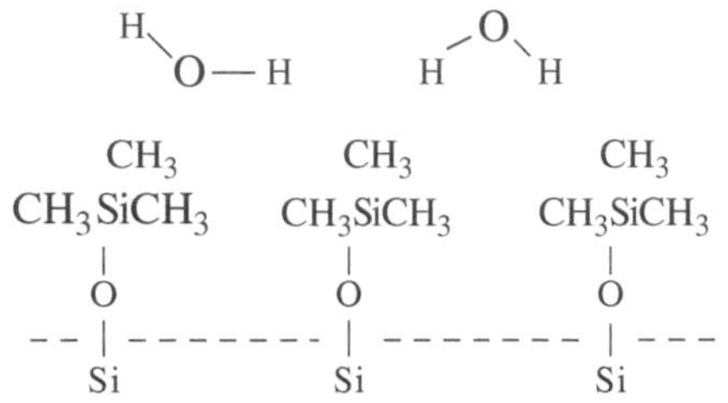

(Figure 1.4 structure)

CH₃   CH₃   CH₃
CH₃SiCH₃   CH₃SiCH₃   CH₃SiCH₃
| | |
O   O   O
-- | ------- | ------- | ---
Si   Si   Si

Figure 1.4 Water molecules can only weakly interact (by vdw forces) with a methylated glass surface.

WETTING PROPERTIES AND THEIR INDUSTRIAL IMPORTANCE

These methylated groups cannot hydrogen bond, and hence water now does not wet and instead forms beads of high 'contact angle' (θ) droplets, and the glass now appears to be hydrophobic, with water droplets similar to those observed on paraffin wax.

This dramatic macroscopic difference in wetting behaviour is caused by only a thin molecular layer on the surface of glass and clearly demonstrates the importance of surface properties. The same type of effect occurs

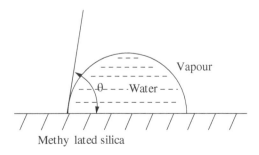

Figure 1.5 A non-wetting water droplet on the surface of methylated, hydrophobic silica.

Figure 1.6 Clean glass flask with water wetting film and start of mist layer intrusion.

every day, when dirty fingers transfer natural grease and fats on to a drinking glass! Thorough cleaning of a glass flask (e.g., by washing in concentrated NaOH solution, as used in dishwashers) initially allows only a thin water film to coat (wet) the inside of the flask, producing visible coloured interference patterns (see photo). When contaminants from the air atmosphere slowly leak in via the stopper, this wetting film is displaced by a fine layer of non-wetting droplets, forming an opaque mist film.

Surface treatments offer a remarkably efficient method for the control of macroscopic properties of materials. When insecticides are sprayed onto plant leaves, it is vital that the liquid wet and spread over the surface, rather than form a mist layer. Another important example is the froth flotation technique, used by industry to separate about a billion tons of ore each year. Whether valuable mineral particles will attach to rising bubbles and be 'collected' in the flotation process is determined entirely by the surface properties or surface chemistry of the mineral particle, and this can be controlled by the use of low levels of 'surface-active' materials, which will selectively adsorb and change the surface properties of the mineral particles. Very large quantities of minerals are separated simply by the adjustment of their surface properties.

Although it is relatively easy to understand why some of the macroscopic properties of liquids, especially their shape, can depend on surface properties, it is not so obvious for solids. However, the strength of a solid is determined by the ease with which micro-cracks propagate when placed under stress, and this depends on its surface energy. This understanding is crucial for the safe design of aircraft. That is, the design must consider the amount of (surface) work required to continue a crack and hence expose new surface. This also has the direct effect that materials are stronger in a vacuum, where their surface energy is not reduced by the adsorption of either gases or liquids typically available under atmospheric conditions. Much lighter structures can be made in space.

Many other industrial examples where colloid and surface chemistry plays a significant role will be discussed later; these include:

- latex paint technology
- water treatment
- cavitation
- emulsions and microemulsions
- soil science
- soaps and detergents
- food science
- mineral processing

Recommended Resource Books:

Adamson, A. W. (1990)	*Physical Chemistry of Surfaces*, 5th edn, New York, Wiley.
Birdi, K. S. (ed,) (1997)	*CRC Handbook of Surface and Colloid Chemistry*, Boca Raton, FL: CRC Press.
Evans, D. F. and H. Wennerstrom (1999)	*The Colloidal Domain*, 2nd edn, New York: Wiley.
Hiemenz, P.C (1997)	*Principles of Colloid and Surface Chemistry*, 3rd edn, New York: Marcel Dekker.
Hunter, R. J. (1987)	*Foundations of Colloid Science*, Vol. 1, Oxford: Clarendon Press.
Hunter, R. J. (1993)	*Introduction to Modern Colloid Science*, Oxford: Oxford Sci. Publ.
Israelachvili, J, N. (1985)	*Intermolecular and Surface Forces*, London: Academic Press.
Ninham, B. W. and P. Lo Nostro (2010)	*Molecular Forces and Self Assembly*, Cambridge: Cambridge University Press.
Shaw, D. J. (1992)	*Introduction to Colloid and Surface Chemistry*, 4th edn, Oxford, Boston: Butterworth-Heinemann

Appendix

A. DISPERSED PARTICLE SIZES

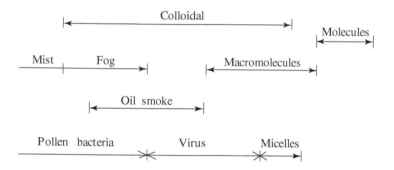

B. SOME HISTORICAL NOTES ON COLLOID AND SURFACE CHEMISTRY

John Freind at Oxford (1675–1728) was the first person to realise that intermolecular forces are of shorter range than gravity.

Young (1805) estimated range of intermolecular forces at about 0.2 nm. Turns out to be something of an underestimate.

Young and **Laplace** (1805) derived meniscus curvature equation.

Brown (1827) observed the motion of fine particles in water.

Johannes D. van der Waals (1837–1923) was a schoolmaster who produced a doctoral thesis on the effects of intermolecular forces on the properties of gases (1873).

Michael Faraday (1857) made colloidal solutions of gold.

Thomas Graham (1860) recognised the existence of colloids in the mid-19th century.

Schulze and Hardy (1882–1900) studied the effects of electrolytes on colloid stability.

Perrin (1903) used the terms 'lyophobic' and 'lyophilic' to denote irreversible and reversible coagulation.

Ostwald (1907) developed the concepts of 'disperse phase' and 'dispersion medium'.

Gouy and Chapman (1910–1913) both independently used the Poisson-Boltzmann equations to describe the diffuse electrical double- layer formed at the interface between a charged surface and an aqueous solution.

Ellis and Powis (1912–1915) introduced the concept of the critical zeta potential for the coagulation of colloidal solutions.

London (1920) first developed a theoretical basis for the origin of intermolecular forces.

Debye (1920) used polarisability of molecules to estimate attractive forces.

Debye and Huckel (1923) used a similar approach to Gouy and Chapman to calculate the activity coefficients of electrolytes.

Stern (1924) introduced the concept of specific ion adsorption at surfaces.

Kallmann and Willstatter (1932) calculated van der Waals force between colloidal particles using summation procedure and suggested that a complete picture of colloid stability could be obtained on the basis of electrostatic double-layer and van der Waals forces.

Bradley (1932) also independently calculated van der Waals forces colloidal particles.

Hamaker (1932) and de Boer (1936) calculated van der Waals forces on macroscopic bodies using summation method.

Derjaguin, Landau, Verwey, and Overbeek (1941–1948) developed the DLVO theory of colloid stability.

Lifshitz (1955–1960) developed a complete quantum electrodynamic (continuum) theory for the van der Waals interaction between macroscopic bodies.

2

Surface Tension and Wetting

The equivalence of the force and energy description of surface tension and surface energy. Derivation of the Laplace pressure and a description of common methods for determining the surface tension of liquids. The surface energy and cohesion of solids, liquid wetting and the liquid contact angle. Laboratory projects for measuring the surface tension of liquids and liquid contact angles.

THE EQUIVALENCE OF THE FORCE AND ENERGY DESCRIPTION OF SURFACE TENSION AND SURFACE ENERGY

It is easy to demonstrate that the surface energy of a liquid actually gives rise to a 'surface tension' or force acting to oppose any increase in surface area. Thus, we have to 'blow' to create a soap bubble by stretching a soap film. A spherical soap bubble is formed in response to the tension in the bubble surface.

Applied Colloid and Surface Chemistry, Second Edition. Richard M. Pashley and Marilyn E. Karaman.
© 2021 John Wiley & Sons Ltd. Published 2021 by John Wiley & Sons Ltd.
Companion website: www.wiley.com/go/pashley/appliedcolloid2e

Figure 2.1   Photograph of a soap bubble.

The soap film shows interference colours at the upper surface, where the film is starting to thin, under the action of gravity, to thicknesses of the order of the wavelength of light. Some beautiful photographs of various types of soap films are given in *The Science of Soap Films and Soap Bubbles* by C. Isenberg (Dover, 1992). Below is an example of the complex coloured images which can be produced by a simple stretched soap film through the interference of light waves:

Figure 2.2   Photograph of flat soap film with a variety of drainage thicknesses giving different colours.

If we stretch such a soap film on a wire frame, we find that we need to apply a significant, measurable force F, to prevent collapse of the film:

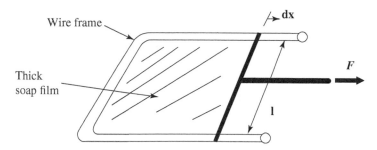

Figure 2.3   Diagram of a soap film stretched on a wire frame.

The magnitude of this force can be obtained by consideration of the energy change involved in an infinitesimal movement of the cross-bar by a distance dx, which can be achieved by doing reversible work on the system, thus raising its free energy by a small amount Fdx. If the system is at equilibrium, this change in (free) energy must be exactly equal to the increase in surface (free) energy $(2dxl\gamma)$ associated with increasing the area of both surfaces of the soap film. Hence, at equilibrium:

$$Fdx = 2dxl\gamma \qquad (2.1)$$

or

$$\gamma = F/2l \qquad (2.2)$$

It is precisely this, that work has to be done to increase a liquid's surface area, which makes the surface of a liquid behave like a stretched skin, hence the term 'surface tension'. It is this tension that allows a 'water boatman' insect to travel freely on the surface of a stagnant pond, locally deforming the skin-like surface of the water.

This simple experimental system clearly demonstrates the equivalence of surface energy and tension. The dimensions of surface energy, mJm^{-2}, are equivalent to those of surface tension, mNm^{-1}. For pure water, an energy of about 73 mJ is required to create 1 m² area of new surface. Assuming that 1 water molecule occupies an area of roughly 12 Å², the free energy of transfer of one molecule of water from bulk to the surface is about 3 kT (i.e. 1.2×10^{-20} J), which compares with roughly 8 kT per hydrogen bond. The energy or work required to create new water-air surface is so crucial to a new-born baby that nature has developed lung surfactants specially designed to reduce this work by about a factor of three. Premature babies often lack this surfactant, and it has to be sprayed into their lungs to help them breathe.

DERIVATION OF THE LAPLACE PRESSURE EQUATION

Since it is relatively easy to transfer molecules from bulk liquid to the surface (e.g. shake or break up a droplet of water), the work done in this process can be measured, and hence we can obtain the value of the surface energy of the liquid. Measuring the surface energy of solids is obviously going to be more difficult (see the later section). The diverse methods for measuring surface and interfacial energies of liquids generally depend on measuring either the pressure difference across a curved interface or the equilibrium (reversible) force required to extend the area of a surface, as above. The former method uses a fundamental equation for the pressure generated across any curved interface; this is the Laplace equation, which is derived in the following section.

Let us consider the conditions under which an air bubble (i.e. a curved surface) is stable. Consider the case of an air bubble produced in water by blowing through a tube, as illustrated below. Note that in this case there is only one surface, unlike the two surfaces produced by soap films.

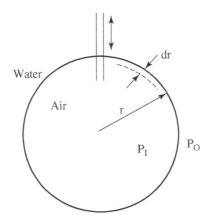

Figure 2.4   Diagram of a spherical air bubble in water.

Obviously, to blow the air bubble we must have applied a higher pressure, P_I, inside the bubble, compared with the external pressure in the surrounding water (P_O). The bubble will be stable when there is no net air flow, in or out, and the bubble radius stays constant. Under these equilibrium conditions, there will be no free energy change in the system for any infinitesimal change in the bubble radius, that is $dG/dr = 0$, where dr is an infinitesimal decrease in bubble radius. If the bubble were to collapse by a small amount dr, the surface area of the bubble will be reduced, giving a decrease in the surface free energy of the system. The only mechanism by which this change can be prevented is to raise the pressure inside the bubble so that $P_I > P_O$ and work has to be done to reduce the bubble size. The bubble will be precisely at equilibrium when the change in free energy due to a reduced surface area is balanced by this work. For an infinitesimal change dr, the corresponding free energy change of this system is given by the sum of the decrease in surface free energy and the mechanical work done against the pressure difference across the bubble surface, thus:

$$dG = -\gamma \left\{ 4\pi r^2 - 4\pi (r - dr)^2 \right\} + (P_I - P_O) 4\pi r^2 dr \tag{2.3}$$

$$= -8\pi r dr \gamma + \Delta P 4\pi r^2 dr \tag{2.4}$$

(ignoring higher-order terms)

At equilibrium $dG/dr = 0$, and hence:

$$-8\pi r \gamma + \Delta P 4\pi r^2 = 0 \tag{2.5}$$

$$\therefore \Delta P = \frac{2\gamma}{r} \tag{2.6}$$

This result is the Laplace equation for a single spherical interface. In general, that is for any curved interface, this relationship expands to include the two principal radii of curvature, R_1 and R_2:

$$\Delta P = \gamma (1/R_1 + 1/R_2) \tag{2.7}$$

Note that for a spherical surface $R_1 = R_2 = r$ and we again obtain equation (2.6). This equation is sometimes referred to as the Young-Laplace equation. The work required to stretch the rubber of a balloon is directly analogous to the interfacial tension of the liquid surface. That the pressure inside a curved meniscus must be greater than that outside is most easily understood for gas bubbles (and balloons) but is equally valid for liquid droplets. The Laplace equation is also useful in calculating the initial pressure required to nucleate very small bubbles in liquids. Very high internal pressures are required to nucleate small bubbles, and this remains an issue for de-gassing, boiling and decompression sickness. Some typical values for bubbles in water are given below:

Bubble radius/nm	1	2	10	1000
Laplace pressure/bar	1440	720	144	1.44

The high pressures associated with high curvature interfaces lead directly to the use of boiling chips to help nucleate bubbles with lower curvature using the porous, angular nature of the chips, as illustrated below.

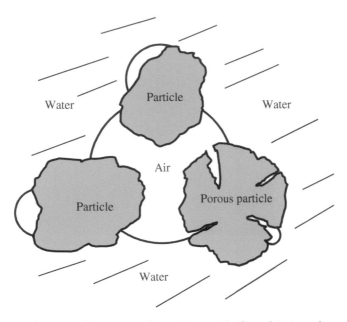

Figure 2.5   Schematic illustration of particles (e.g. boiling chips) used to reduce air bubble curvature.

Control of the pressure in a liquid hydrogen bubble chamber is also used to control bubble sizes for tracking subatomic particles produced in a high energy particle accelerator.

METHODS FOR DETERMINING THE SURFACE TENSION OF LIQUIDS

The equilibrium curvature of a liquid surface or meniscus depends not just on its surface tension but also on its density and the effect of gravity. The variation in curvature of a meniscus surface must be due to hydrostatic pressure differences at different vertical points on the meniscus. If the curvature at a given starting point on a surface is known, the adjacent curvature can be obtained from the Laplace equation and its change in hydrostatic pressure $\Delta h \rho g$.

In practice the liquid droplet, say in air, has a constant volume and is physically constrained at some point, for example, when a pendant drop is constrained by the edge of a capillary tube.

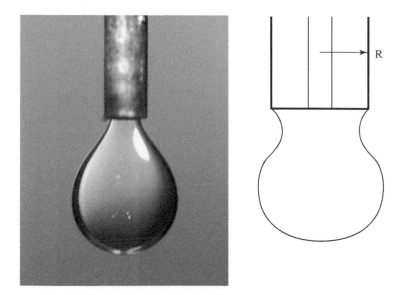

Figure 2.6   Photograph and diagram of a pendant liquid drop in air at the end of a capillary tube.

For given values of the total volume, the radius of the tube R, the density ρ and the surface energy γ, the shape of the droplet is completely defined and can be calculated using numerical methods (e.g. the Runge-Kutta method) to solve the Laplace equation. Beautiful shapes can be generated using this numerical procedure, as illustrated in the following shape calculated theoretically using the Laplace equation. Although a wide variety of shapes can be generated using the Laplace equation in a gravitation field, only those shapes which give a minimum in the total energy (that is, surface and potential) will be physically possible.In practice, a continuous series of numerically generated profiles is calculated until the minimum energy shape is obtained. An example of a theoretically generated droplet shape is shown in the following graph. From these profile curves it is easy to generate the corresponding volume of the droplet.

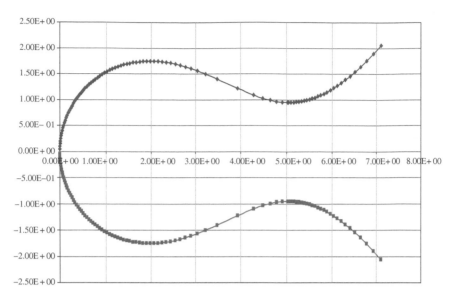

Figure 2.7 Example of a theoretical calculation of the shape of a liquid droplet, suspended under gravity, obtained using the Laplace equation.

It is interesting to consider the size of droplets for which surface (tension) forces, compared with gravity, dominate liquid shapes. A simple balance of these forces is given by the relation:

$$length = \sqrt{\frac{\gamma}{\rho g}} \approx 4mm, \quad \text{for water.}$$

Thus we would expect water to 'climb' up the walls of a clean (i.e. water wetting) glass vessel for a few millimetres but not more, and we would expect a sessile water droplet to reach a height of several millimetres on a hydrophobic surface before the droplet surface is flattened by gravitational forces. The curved liquid border at the perimeter of a liquid surface or film is called the Plateau border after the French scientist who studied liquid shapes after the onset of blindness following his personal experiments on the effects of sunlight on the human eye. The observation of a pendant drop is one of the best methods of measuring surface and interfacial energies of liquids. Either the drop can be photographed and the profile digitised, or published tables can be used to obtain γ from only the drop volume and the minimum and maximum width of the drop. Another simple method of measuring the surface energy of liquids is using a capillary tube. In this method, the height to which the liquid rises in the capillary above the free liquid surface is measured. This situation is illustrated below:

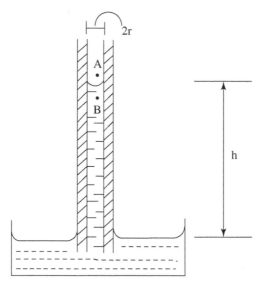

Figure 2.8   Schematic diagram of the rise of a liquid that wets the inside walls of a capillary tube.

Using the Laplace equation, the pressure difference between points A and B is given simply by $\Delta P = 2\gamma/r$, if we assume that the meniscus is hemispherical and of radius r. However, this will be accurate only if the liquid wets the walls of the glass tube. If the liquid has a finite contact angle θ with the glass, as illustrated below,

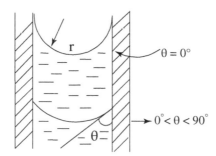

Figure 2.9   Schematic diagram of the shape of the meniscus for complete wetting and non-wetting liquids.

then from simple geometry (again assuming the meniscus is spherical):

$$\Delta P = \frac{2\gamma \cos\theta}{r} \tag{2.8}$$

Note that if $\theta > 90°$ (e.g. mercury on glass), the liquid will actually fall below the reservoir level, since the meniscus will be curved in the opposite direction.

The pressure difference between parts A and B must be equal to the hydrostatic pressure difference $h\rho g$ (where ρ is the density of the liquid and the density of air is ignored). Thus, we obtain the result that:

$$h\rho g = \frac{2\gamma \cos\theta}{r} \tag{2.9}$$

and hence

$$\gamma = \frac{rh\rho g}{2\cos\theta} \tag{2.10}$$

from which measurement of the capillary rise and the contact angle gives the surface tension of the liquid (the factors that determine the contact angle will be discussed in the following section). Although equation (2.10) was derived directly from the mechanical equilibrium condition which must exist across any curved interface, this is not the reason why the liquid rises in the capillary. This phenomenon occurs because the interfacial energy of the clean glass/ water interface is much lower than that of the glass/air interface. The amount of energy released on wetting the glass surface and the potential energy gained by the liquid on rising in a gravitational field must be minimized at equilibrium. Equation (2.10) can, in fact, be derived from this (free energy minimization) approach, shown below. It is also interesting to note that because these interfacial energies are due to short-range forces, that is, surface properties, the capillary walls could be as thin as 100 Å, and the liquid would still rise to exactly the same height (compare this with the gravitational force).

CAPILLARY RISE AND A FREE ENERGY ANALYSIS

The fundamental reason why a liquid will rise in a narrow capillary tube against gravity must be that $\gamma_{sv} > \gamma_{sl}$, i.e. that the free (surface) energy reduction on wetting the solid is balanced by the gain in gravitational potential energy. The liquid will rise to a height h at which these factors are balanced. Thus, we must find the value of h for the equilibrium condition: $\frac{dG_T}{dh} = 0$, where G_T is the total free energy of the system at constant temperature. For a given height h:

$$\text{potential energy (increase)} = \pi r^2 h\rho g \frac{h}{2} \quad (i.e. \, centre \, of \, gravity \, at \, h/2)$$

$$\text{surface energy (decrease)} = 2\pi rh(\gamma_{SV} - \gamma_{SL})$$

$$\therefore \frac{d\left\{\pi r^2 h^2 \rho g / 2 - 2\pi r h\left(\gamma_{SV} - \gamma_{SL}\right)\right\}}{dh} = 0, \; at \, h = h_{equil.}$$

$$i.e. \; \pi r^2 \rho g h - 2\pi r\left(\gamma_{SV} - \gamma_{SL}\right) = 0$$

$$\therefore \gamma_{SV} - \gamma_{SL} = \frac{r\rho g h}{2}$$

and since $\gamma_{SV} = \gamma_{SL} + \gamma_{LV} \cos\theta$ (Young equation, see later)

$$\therefore h = \frac{2\gamma_{LV}\cos\theta}{r\rho g}, \; as \, before.$$

The capillary rise method, although simple, is in practice not as useful as the pendant drop method because of several experimental problems, such as the need to determine the contact angle, non-sphericity of the meniscus and uneven bore of the capillary.

One industrial application of the Laplace pressure generated in a pore is the use of Goretex membranes (Teflon porous membranes) to concentrate orange juice and other juices to reduce their bulk and hence export costs. This process depends on the Laplace pressure retaining vapour in the Teflon pores to allow water to be drawn through them as vapour into a concentrated salt solution on the other side of the membrane. As can be seen from the simple calculation illustrated below, as long as the water contact angle remains high, say at around 110°, the pressure required to push water into the pores is greater than the hydrostatic pressure used in the operation and the juice can be successfully concentrated. Unfortunately, this process is very sensitive to the presence of surface active ingredients in the juice, which can

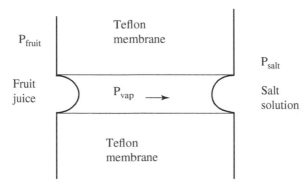

Figure 2.10 Schematic diagram of the concentration of fruit juice via water vapour transport across a porous Teflon membrane.

reduce the contact angle, allowing the pores to become filled with water and the juice become contaminated with salt. This process is illustrated below, for which the Laplace pressures generated depend on the contact angle of water on the Teflon surface:

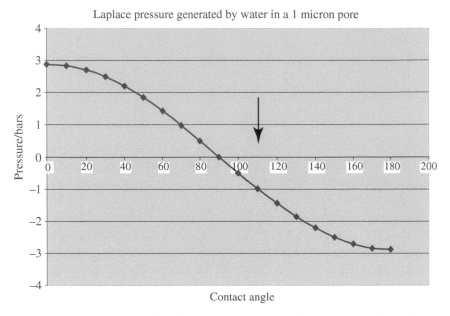

Figure 2.11   The calculated Laplace pressure generated across a curved interface as a function of contact angle.

The dramatic effect of Laplace pressure can also be easily demonstrated using a syringe filled with water and attached to a Teflon micron-sized membrane. Water cannot be pushed into the membrane; however, simply wetting the membrane with a droplet of ethanol will fill the pores, and then the syringe easily pushes water through the membrane.

THE KELVIN EQUATION

It is often also important to consider the pressure of the vapour in equilibrium with a liquid. It can be demonstrated that this pressure, at a given temperature, actually depends on the curvature of the liquid interface. This follows from the basic equations of thermodynamics, given in Chapter 4, which lead to the result that:

$$\partial \mu = +V_m \partial P$$

That is, the chemical potential of a component increases linearly with the total pressure of the system. (V_m is the partial molar volume of the component.) Thus, if we consider the change in chemical potential of the vapour and the liquid on producing a curved surface:

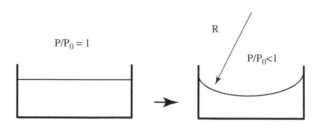

Figure 2.12 Schematic diagram showing that the equilibrium vapour pressure changes with the curvature of the liquid-vapour interface.

It follows that the change in chemical potential of the vapour is given by:

$$\Delta\mu^v = \mu^v - \mu^0 = RT\ln_e\left(\frac{P}{P_o}\right)$$

Now, since both cases are at equilibrium, there must be an equivalent decrease in chemical potential of the liquid i,e.:

$$\Delta\mu^v = \Delta\mu^l$$

But from the Laplace equation, the change in pressure of the liquid (assuming the meniscus is, for simplicity, spherical) is given by:

$$\Delta P = \frac{2\gamma}{r_e}$$

where r_e is the equilibrium radius of the (spherical) meniscus. Thus, it follows that the change in chemical potential of the liquid must be given by:

$$\Delta\mu_l = V_m\Delta P = \frac{2\gamma V_m}{r_e}$$

where combining with the earlier equation for the change in chemical potential of the vapour gives the result:

$$\Delta\mu^{v} = \mu^{v} - \mu^{0} = RT\ln_{e}\left(\frac{P}{P_{o}}\right) = \Delta\mu_{l} = \frac{2\gamma V_{m}}{r_{e}}$$

which on rearrangement gives the Kelvin equation for spherical menisci:

$$RT\ln_{e}\left(\frac{P}{P_{0}}\right) = \frac{2\gamma V_{m}}{r_{e}} \quad \text{or, in general:} \quad RT\ln_{e}\left(\frac{P}{P_{0}}\right) = \gamma V_{m}\left(\frac{1}{R_{1}} + \frac{1}{R_{2}}\right)$$

This relationship gives some interesting and useful predictions for the behaviour of curved interfaces. For example, water at P/P_{0} values of 0.99 should condense in cracks or capillaries and produce menisci of (negative) radius 105 nm, of the type shown below:

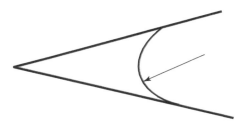

Figure 2.13   Capillary condensation of water vapour into a crack.

whereas for a sessile droplet, there must be a positive Kelvin radius, and for typical large droplets of, say, mm radius, they must be in equilibrium with vapour very close to saturation.

Figure 2.14   Diagram of a sessile droplet.

A range of calculated values for water menisci at 21 °C are given in the following figure for both concave (negative radius) and convex (positive radius) menisci.

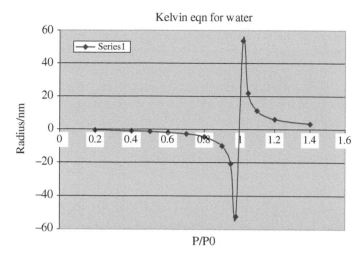

Figure 2.15   Graph of the relative vapour pressure against radius of the correspond-ing equilibrium meniscus. Positive radii are for convex and negative for concave menisci.

Another common method used to measure the surface tension of liquids is called the Wilhelmy plate. This method uses the force (or tension) associ-ated with a meniscus surface to measure the surface energy rather than using the Laplace pressure equation. (Note that in real cases *both* factors usually arise, but often only one is needed to obtain a value for γ.) The Wilhelmy plate is illustrated below:

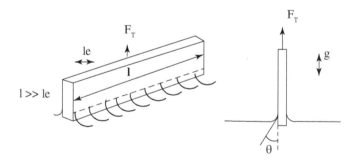

Figure 2.16   Diagram of the Wilhelmy plate method for measuring the surface tension of liquids.

The total force F_T (measured using a balance) is given by:

$$F_T = F_W + 2l\gamma cos\theta \qquad (2.11)$$

where F_w is the dry weight of the plate. (Note that the base of the plate is at the same level as the liquid thereby removing any buoyancy forces.) The plates are normally made of thin platinum which can be easily cleaned in a flame and for which l_e can be ignored. Again, this method has the problem that θ must be known if it is greater than zero. In the related du Noüy ring method, the plate is replaced by an open metal wire ring. At the end of this chapter, a laboratory class is used to demonstrate yet another method which does not require knowledge of the contact angle and involves withdrawal of a solid cylinder attached to a liquid surface.

THE SURFACE ENERGY AND COHESION OF SOLIDS

Measurement of the surface energy of a liquid is relatively easy to both perform and understand. All methods are based on measuring the work required to create a new surface by transferring molecules from bulk liquid. However, what about the surface energy of a solid? Clearly, for solids it is impractical to move molecules from bulk to the surface. There are basically two ways by which we can attempt to obtain the surface energy of solids:

(1) by measuring the cohesion of the solid and,
(2) by studying the wetting behaviour of a range of liquids with different surface tensions on the solid surface.

Both methods are not straightforward, and the results are not as clear as those obtained for liquids. The cohesive energy per unit area W_c is equal to the work required to separate a solid in the ideal process illustrated below. In this ideal process the work of cohesion, W_c, must be equal to twice the surface energy of the solid, γ_s. Although this appears simple as a thought experiment, in practice, it is difficult. For example, we might measure the critical force (F_c) required to separate the material but then we need a theory to relate this to the total work done. The molecules near the surface of the freshly cleaved solid will rearrange after measuring F_c. Also, the new area will not usually be smooth, and hence the true area is much larger than the geometric area. Only a few materials can be successfully studied in this way. One of them is the layered, natural aluminosilicate crystal, muscovite mica, which is available in large crystals and can be cleaved in a controlled manner to produce two molecularly smooth new surfaces.

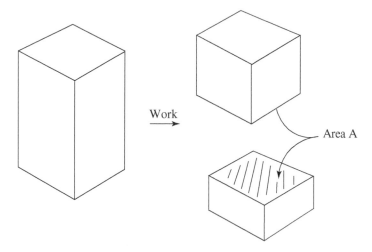

Figure 2.17 Ideal experiment designed to measure the work required to create new area and hence find the value of the surface energy of a solid.

In comparison, the adhesive energy per unit area W_a between two different solids is given by:

$$W_a = \gamma_A + \gamma_B - \gamma_{AB} \tag{2.12}$$

where γ_A and γ_B are the surface energies of the solids and γ_{AB} is the interfacial energy of the two solids in contact (cf. $\gamma_{AA} = 0$). Again the adhesive energy is a difficult property to measure. It is also very hard to find the actual contact area between two different materials, since this is almost always much less than the geometric area. That this is the case is the reason why simply pressing two solids together does not produce adhesion (except for molecularly smooth crystals like mica), and a glue must be used to dramatically increase the contact area. The main function of a glue is to facilitate intimate molecular contact between two solids, so that strong short-range van der Waals forces can hold the materials together.

THE CONTACT ANGLE

The second approach to obtaining the surface energies of solids involves the study of wetting and non-wetting liquids on a smooth clean solid substrate. An example below is of a water droplet on Teflon.

Figure 2.18   An advancing water droplet on Teflon.

Let us examine the situation for a non-wetting liquid (where $\theta > 0°$), which will form a sessile drop on the surface of a solid:

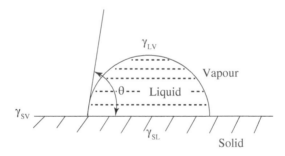

Figure 2.19   Diagram of a sessile droplet.

Using an optical microscope, it is possible to observe and measure a finite contact angle (θ) as the liquid interface approaches the three-phase contact perimeter of the drop. Let us consider the local equilibrium situation along a small length of the three-phase line or TPL. This is the line where all three phases are in contact. Let us examine this region in more detail in the schematic diagram below:

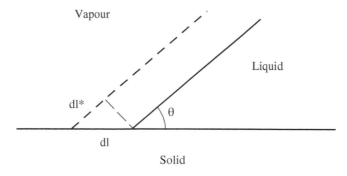

Figure 2.20 Diagram of the three-phase line and its perturbation to determine the contact angle.

Let us examine the equilibrium contact angle, θ, for which an infinitesimal movement in the TPL by distance dl to the left hand side, will not change the total surface free energy of the system. We can consider area changes for each of the three interfaces for unit length 'l' vertical to the page and along the TPL. Thus the total interfacial energy change must be given by the sum:

$$dG = \gamma_{sl}ldl + \gamma_{lv}ldl* - \gamma_{sv}ldl$$

From simple geometry, $dl* = dl\cos\theta$, and hence at equilibrium, where $dG/dl = 0$, it follows that:

$$\gamma_{SV} = \gamma_{SL} + \gamma_{LV}\cos\theta \qquad (2.13)$$

This important result is called the Young equation. It can also be derived by simply considering the horizontal resolution of the 3 surface tensions (i.e. as forces per unit distance), via standard vector addition:

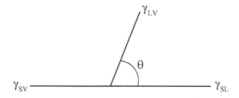

Figure 2.21 Balance of surface energies at the TPL gives the Young equation directly.

However, what becomes of the vertical component? This force is actually balanced by the stresses in the solid around the drop perimeter (or TPL), which can actually be visually observed on a deformable substrate, such as paraffin wax.

Since we can measure the liquid surface energy, γ_{LV}, the value of $(\gamma_{SV} - \gamma_{SL})$ can be obtained, but unfortunately γ_{SL} is as difficult to measure directly as γ_{SV}. However, if θ is measured for a range of liquids with different surface energies, then a plot of $cos\,\theta$ vs γ_{LV} gives a critical surface energy value, γ_c, at $\theta = 0°$ (the complete wetting case). It is often not unreasonable to equate γ_c with γ_{SV}, because in many cases at complete wetting γ_{SL} approaches zero. The following schematic figure corresponds to the type of behaviour observed for a range of different liquids wetting Teflon. The low surface energy of Teflon has been estimated from this type of data.

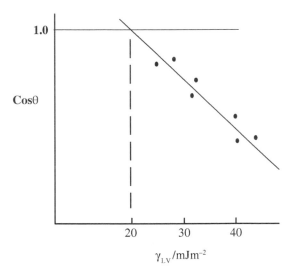

Figure 2.22 Typical plot of the measured contact angles of a range of liquids with different surface tensions, on a low energy solid.

Clearly the surface energy of a solid is closely related to its cohesive strength. The higher the surface energy, the higher its cohesion. This has some obvious and very important ramifications. For example, the strength of a covalently bonded solid, such as a glass or metal, must always be greatest in a high vacuum, where creation of a new surface must require the greatest work. The strength of the same material in water vapour or immersed in liquid water will be much reduced, often by at least an order of magnitude. This is because the freshly formed solid surface must initially be composed of high energy atoms and molecules produced by the cleavage of many chemical bonds. These new high energy surfaces will rapidly adsorb and react with any impingent gas molecules. Many construction materials under strain will therefore behave differently, depending on the environment. It should also be noted that the scoring of a glass rod only goes to a depth of about 0.01% of the rod's thickness but this still substantially reduces its strength. Clearly, crack propagation determines the ultimate strength of any material, and in general cracks will propagate more easily in an adsorbing environment (e.g. of liquid or vapour). Objects in outer space can, therefore, be produced using thinner materials but still with the same strength.

Table 2.1

	θ_A
Paraffin wax	110°
PTFE (teflon)	108°
Polyethylene	95°
Graphite	86°
AgI	70°
Mica	7°
Gold (clean)	0°
Glass (clean)	0°
(Hg on glass	135°)

A list of (advancing) water contact angles on various solid substrates is given in the Table.

It is immediately obvious that water will not wet low energy surfaces (γ_{SV} < 70 mJm^{-2}) such as hydrocarbons, where there is no possibility of either hydrogen bonding or dipole-dipole interactions with the solid substrate. However, complete wetting occurs on high-energy surfaces (γ_{SV} > 70 mJm^{-2}), such as clean glass and most metals. Directly from our concept of surface energies, it is clear that we would expect a liquid to spread on a substrate if:

$$\gamma_{SV} - \gamma_{LV} - \gamma_{SL} > 0 \qquad (2.14)$$

In fact we can define a parameter,

$$S_{LS} = \gamma_{SV} - \gamma_{LV} - \gamma_{SL} \qquad (2.15)$$

called the spreading coefficient. If $S_{LS} > 0$, the liquid will spread. Since γ_{LV} for mercury is about 485 mJm^{-2} (because of very strong metallic bonding), we would not expect this liquid to spread on anything but very high energy surfaces (such as other metals).

The contact angle is usually quite easy to measure and is a very useful indicator for the wetting properties of a material. Whether water will wet a solid or not is used extensively by the mining industry in a separation technique called froth flotation. As mentioned previously, a rock containing a required mineral together with unwanted solid, often quartz, can be ground to give colloidal-size particles of the pure mineral. If the powder is dispersed in water and bubbles are continuously passed through the chamber, those particles with water contact angles greater than about 20° will attach to the rising bubbles, whereas the wetting particles will remain dispersed:

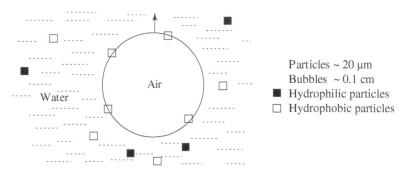

Particles ~ 20 μm
Bubbles ~ 0.1 cm
■ Hydrophilic particles
□ Hydrophobic particles

Figure 2.23   Diagram of a bubble selectively collecting hydrophobic particles in the froth flotation process.

In water the wetted solid is termed 'hydrophilic', whereas the non-wetted solid is 'hydrophobic'. Naturally hydrophobic minerals, such as some types of coal, talc and molybdenite, are easily separated from the unwanted hydrophilic quartz sand (referred to as gangue). However, surfactants and oils are usually added as collectors. These compounds adsorb onto the hydrophilic mineral surface and make it hydrophobic. Frothers are also added to stabilise the foam at the top of the chamber, so that the enriched mineral can be continuously scooped off. The selective flotation of a required mineral depends critically on surface properties, and these can be carefully controlled using a wide range of additives. Throughout the world a large quantity (about 10^9 tons annually) of minerals are separated by this method.

Although the contact angle is a very useful indicator of the energy of a surface, it is also affected by a substrate's surface roughness and chemical microheterogeneity. This can be well illustrated by comparing ideal calculated contact angle values with measured ones. For example, we can easily calculate the expected water contact angle on a liquid or solid pure hydrocarbon surface by using the surface tension of water and hexadecane (27.5 mNm^{-1}) and the interfacial tension between the oil and water (53.8 mNm^{-1}). Use of these values in the Young equation gives an expected angle of 111°, in close agreement with the values observed for paraffin wax. However, on many real surfaces the observed angle is hysteretic, giving quite different values depending on whether the liquid droplet size is increasing (giving the advancing angle θ_A) or decreasing (giving the receding angle θ_R). The angles can differ by as much as 60°, and there is some controversy as to which angle should be used (for example in the Young equation) or even if the average value should be taken. Both θ_A and θ_R must be measured carefully, whilst the three-phase line is stationary but just on the point of moving, either forwards or back. In general, both angles and the differences between the two give indirect information about the state of the surface, and both should always be reported. The degree of hysteresis observed is a measure of both surface roughness and surface chemical heterogeneity.

Industrial Report

Photographic quality printing.

Modern photographic-quality inkjet papers have a surface coating comprising either a thin polymer film or a fine porous layer. In either case the material is formed using a high speed coating process. This process requires careful control to obtain the necessary uniformity together with low manufacturing costs. From a scientific point of view, all coating processes have at least one static wetting line and one dynamic wetting line, and in many cases the behaviour of both the dynamic and static wetting lines determines the speed and uniformity of the final coating. At a fundamental level, surface forces control wetting line behaviour. However, the details of the physical and chemical behaviour of a dynamic wetting line are still a matter of scientific debate.Having created a thin film coating, the liquid sheet is susceptible to a variety of potential defects ranging from holes generated through Marangoni driven flows to mottle driven by interface instabilities. These flow defects can [be], and are, controlled by the use of appropriate surfactant additives.

Porous inkjet papers are in general created from colloidal dispersions. The eventual random packing of the colloid particles in the coated and dried film creates an open porous structure. It is this open structure that gives photographic-quality inkjet paper its 'apparently dry' quality as it comes off the printer. Both the pore structure and pore wettability control the liquid invasion of the coated layer and therefore the final destination of dyes. Dispersion and stability of the colloidal system may require dispersant chemistries specific to the particle and solution composition. In many colloidal systems particle-particle interactions lead to flocculation, which in turn leads to an increase in viscosity of the system. The viscosity directly influences the coating process through the inverse relation between viscosity and maximum coating speed.

Surface and colloid science can also play a significant role in formulation of pigmented inks, another colloidal dispersion again requiring a good dispersant for stability within the ink cartridge. Jetability of the ink from the printhead and the wetting behaviour of the ink on and in the paper are both controlled by surface interactions. Inkjet material manufacture and design therefore provides a fertile ground for the surface and colloid scientist to apply his skills.

Dr Andrew Clarke
Surface and Colloid Science Group
Research & Development, Kodak Limited
Harrow, UK

SAMPLE PROBLEMS

(1) What pressure must be applied to force water through an initially dry Teflon membrane which has a uniform pore size of 0.5 μm diameter? What factors can reduce this pressure? Give examples of the industrial and everyday use of this type of porous material. ($\theta_{water} = 110°$, $\gamma_{water} = 73$ mNm^{-1}.)

(2) Use the Laplace equation to calculate the spherical radius of the soap film which is formed by the contact of two bubbles with radii of 1 and 3 cm. Assume that the soap bubbles have a surface tension of 30 mJm^{-2}. Draw a sketch of the contacting bubbles to help you.

(3) Calculate the surface tension of a wetting liquid, of density 1.2 g/ml, which rises 1.05 m in a vertical capillary tube with an internal diameter of 0.2 mm.

(4) Two curved regions on the surface of a water droplet have principal radii of 0.2, 0.67 and 0.1, 0.5 cm. What is their difference in vertical height?

(5) Water can just be forced through a Goretex (porous Teflon) membrane at an applied pressure of 1.5 bar. What is the pore size of the membrane?

(6) Explain why it is possible to make sandcastles only in partially wet sand and not in either fully immersed or completely dry sand.

(7) Why do construction workers spray water when they are working with very dusty materials?

(8) The figure below shows two spherical hydrophilic solid particles held together by a water meniscus. If the upper particle is held fixed, calculate the minimum force (F) required to pull the lower particle away. Assume that water has a zero contact angle with the solid.

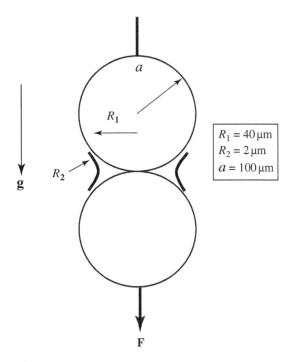

Figure 2.24   Two spherical particles held together by a water meniscus.

(9) A colloidal particle is held on to a solid surface by a water meniscus, as illustrated in the diagram. Estimate the minimum force (F) required to detach the particle. (Ignore the mass of the colloid.)

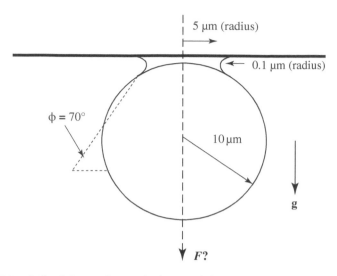

Figure 2.25 Colloidal particle attached to a solid surface by a water meniscus.

Experiment   Rod-In-Free-Surface (Rifs) Method for the Measurement of the Surface Tension of Liquids

Introduction

There is a wide range of different methods used to measure surface tension/energy of liquids and solutions. The RIFS technique has the advantage that it requires only simple, easily available (cheap) equipment and yet gives absolute and accurate surface energies (to 0.1 mJm^{-2}). The principle of this method is that the maximum force on a

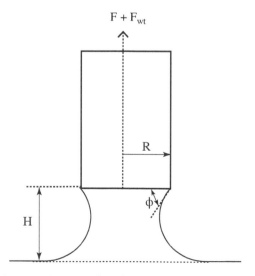

Figure 2.26 Schematic diagram of a cylindrical rod pulled from a liquid surface.

cylindrical rod pulled through the surface of a liquid (as shown in Figure 1) is related to the surface tension of the liquid.

The rod lifts up a meniscus above the force surface of the liquid, and the weight of this meniscus can be measured. The additional force on the rod (i.e. above its own weight) must be given by:

$$F = \pi R^2 H \rho g + 2\pi R \gamma \sin\Phi \qquad (1)$$

The first term on the right-hand side is due to the hydrostatic pressure (or suction) acting over the base of the rod, and the second term is due to surface tension forces around the perimeter. (It should be noted that Φ is not equal to the equilibrium contact angle θ but is determined by the meniscus shape).

If we could accurately photograph the profile illustrated in Figure 1, we could measure H, R and Φ, and since F is simply the excess weight on the rod, obtain γ. However, experimentally this would be difficult, and significant errors are likely in measuring the height H.

A much easier method has been developed by Padday et al., *Faraday Trans. I*, **71**, 1919 (1975), which only requires measurement of the maximum force or weight on the rod as it is pulled upwards. It has been shown by using the Laplace equation to generate meniscus profiles that this maximum is stable and quite separate to the critical pull-off force where the meniscus ruptures. A typical force-height curve is shown in Figure 2.

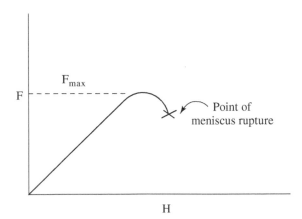

Figure 2.27 Illustration of the force pulling on the rod as a function of height above the liquid surface.

Padday et al. derived an equation which allows the surface tension of the liquid to be obtained from the measured maximum force on the rod, the known density of the liquid (ρ) and the radius of the cylinder. This equation is given below:

$$R/k = a_o + a_1\left(R^3/V\right) + a^2\left(R^3/V\right)^2 + a_3\left(R^3/V\right)^3 \qquad (2)$$

where $k = (\gamma/\rho g)^{0.5}$ and V is the volume of the meniscus raised above the level of the liquid, which is directly obtained from the measured weight of the meniscus. a_o, a_1, a_2 and a_3 are constants which depend on the value of R^3/V and are given in the following table. Once the value of R^3/V is measured, the correct values of a_n are chosen and used to calculate k and hence γ.

R^3/V	a_0	a_1	a_2	a_3
0.01–0.02	9.07578×10^{-2}	2.07380×10^1	-4.46445×10^2	6.23543×10^3
0.02–0.03	1.15108×10^{-1}	1.64345×10^1	-2.00113×10^2	1.63165×10^3
0.03–0.04	1.06273×10^{-1}	1.69246×10^1	-2.04837×10^2	1.57343×10^3
0.04–0.05	6.34298×10^{-2}	1.88348×10^1	-2.20374×10^2	1.43729×10^3
0.05–0.07	1.56342×10^{-1}	1.23019×10^1	-6.96970×10^1	2.93803×10^2
0.07–0.10	2.21619×10^{-1}	9.31363	-2.39480×10^1	5.96204×10^1
0.10–0.15	3.11064×10^{-1}	6.97932	-3.58929	0.0
0.15–0.20	3.67250×10^{-1}	6.26621	-1.32143	0.0
0.20–0.30	4.40580×10^{-1}	5.60569	1.63171×10^{-1}	0.0
0.30–0.40	4.47385×10^{-1}	5.63077	0.0	0.0
0.40–0.50	4.72505×10^{-1}	5.39906	4.24569×10^{-1}	0.0
0.50–0.60	3.780×10^{-1}	5.80000	0.0	0.0
0.60–0.80	5.72110×10^{-1}	5.15631	5.33894×10^{-1}	0.0
0.80–1.00	2.99048×10^{-1}	5.86260	7.83455×10^{-2}	0.0
1.00–1.20	6.76415×10^{-1}	5.16281	4.01204×10^{-1}	0.0
1.20–1.40	4.08687×10^{-2}	6.20312	-2.40752×10^{-2}	0.0
1.40–1.60	2.53174×10^{-1}	5.90351	8.14259×10^{-2}	0.0
1.60–1.85	-1.30000×10^{-2}	6.20000	0.0	0.0

Experimental procedures

This experiment requires skill and careful technique in order to obtain accurate results. As with most surface chemistry experiments, cleanliness is of paramount importance. High-energy liquids such as water easily pick up surface-active contaminants from the air in a laboratory, and great care should be taken to reduce exposure. Contaminants generally do not adsorb at the surface of low-energy liquids such as hexane and hence are less of a problem.

The liquids to be studied in this experiment are water, hexane, n-octanol and aqueous solutions of CTAB. It is recommended that they be measured in the order written, where the most critical with respect to contamination is first. The water used should be the best available, such as double distilled, and should be stored in a sealed flask before use. Pure samples of the other liquids should also be used as well as top quality water to make up the CTAB solutions. The CTAB solutions should be measured at concentrations of 0.01, 0.1, 0.3, 0.6, 1 and 10 mM at a temperature above 30°C. CTAB has a Krafft temperature around 25 °C; below this temperature the surfactant will precipitate from aqueous solution at the higher concentrations (see later).

The apparatus used to measure the maximum force on a rod is shown in Figure 3. Be extremely careful in connecting the rod to the balance hook (the balance should be switched off during this procedure). An electronic balance is usually more robust and easier to use.

The mechanical stage is used to roughly position the level of the liquid; the level is then gradually and carefully lowered by withdrawing liquid into the motor-driven syringe. Assuming that there are no vibrations and the rod is almost perfectly vertical, it should be possible to raise the rod beyond the force maximum without rupturing the meniscus. The maximum weight can then be accurately found by adding and removing liquid in this region. (This must be done very slowly for the CTAB solutions to allow equilibrium adsorption of the surfactant.) The measurement is acceptable only when the meniscus is symmetrical around the perimeter of the base of the rod. The temperature of the liquid should be noted. The maximum weight of the meniscus is equal to the maximum measured weight minus the dry weight of the rod.

The syringe, dish and rod must be thoroughly cleared on changing liquids. For the water measurements, Pyrex glass and the stainless steel rod are best cleaned by (analar) ethanol scrubbing and washing followed

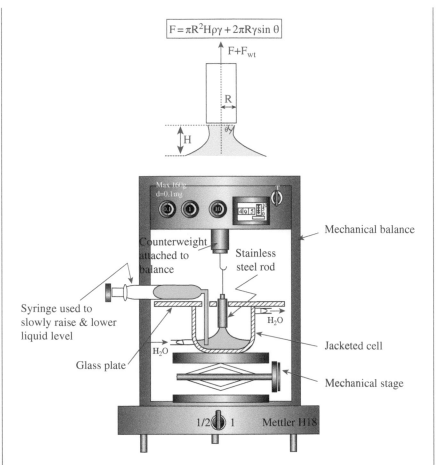

$$F = \pi R^2 H \rho \gamma + 2\pi R \gamma \sin \theta$$

Figure 2.28 Diagram of the apparatus used to measure the surface tension of liquids by the RIFS method.

by rinsing in high-quality water. For the organic liquids, ethanol rinsing is sufficient with final rinsing using the liquid to be measured. (Note that the rod must be free of excess liquid during measurement and also there must be no liquid film on the sides of the rod).

The base diameter of the rod must be accurately measured using a micrometer and taking an average over at least three positions around the base. That this measurement is critical can be seen from an examination of Equation 2. The perimeter of the base must be sharp and undamaged.

Calculate the surface energies of each of these liquids and plot a graph of γ for the CTAB solutions as a function of $\log_{10} (conc.)$. Use your results and the Gibbs adsorption equation (see later) to estimate the minimum surface area per CTAB molecule adsorbed at the air-water interface.

FOR CONSIDERATION/TYPICAL QUESTIONS

(1) What advantages does this (RIFS) technique have over the Wilhelmy plate method?
(2) Why is the surface energy of octanol higher than that of hexane?
(3) What is the cmc of CTAB solutions?
(4) Why do we plot the CTAB concentration as a log function?

Experiment   Contact Angle Measurements

Introduction

The value of the equilibrium contact angle (θ) at the three-phase line (TPL) produced by a liquid droplet placed on a flat, solid substrate is determined by the balance of interfacial energies at each surface. Thomas Young derived an equation describing this situation in 1804:

$$\gamma_{SV} = \gamma_{SL} + \gamma_{LV}\cos\theta \qquad (1)$$

where γ_{SV}, γ_{SL} and γ_{LV} are the energies of the solid/vapour, solid/liquid and liquid/vapour interfaces. The contact angle θ is the angle subtended between the liquid/vapour and liquid/solid interfaces and is illustrated in the following figure.

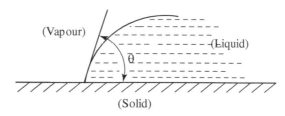

(Vapour)   (Liquid) θ (Solid)

Figure 2.29   Contact angle made by a sessile droplet.

In order for the droplet to be at equilibrium the gas phase must be saturated with the liquid vapour. For this reason, measurements are normally carried out in a sealed vessel with good temperature control.

Contact angle measurements are of fundamental importance in a range of industrial and everyday processes such as flotation, painting (i.e. the paint must wet the substrate) and weatherproofing. In the flotation process a solid block of the powdered mineral to be floated is often studied using a wide range of collector (i.e. surfactant) solutions to determine optimum flotation conditions.

The surface energies of liquids can be directly measured, but this is not the case for most solid/vapour and solid/liquid energies (with one or two notable exceptions, such as mica). An estimate of solid/vapour energies can, however, be obtained by measuring the equilibrium contact angles of a range of different liquids of higher surface energy than the substrate, i.e. non-wetting liquids. A Zisman plot of $\cos\theta$ against γ_{LV} is often linear and can be extrapolated to $\cos\theta = 1$ (i.e. $\theta = 0°$) to give the value of the 'critical surface tension' γ_c of the solid, where a liquid will just completely wet the surface. It has been shown that γ_c is a reasonable approximation for γ_{SV}.

In this experiment, γ_c will be determined for methylated (hydrophobic) soda glass. Both advancing (θ_A) and receding (θ_R) angles will be measured in order to estimate the degree of hysteresis in each case. Use both values of θ_A and θ_R, as well as the average, to produce several Zisman plots.

Experimental procedures

Contact angle measurement

The contact angles can be measured by observing the TPL through a microscope which has a rotating eyepiece and cross hairs. The eyepiece is used to measure the angle of the cross hairs on a protractor scale. By tilting the microscope slightly, the reflected image of the liquid droplet can also be observed, and the double-angle (2θ) can be measured, which increases the accuracy. An illustration of this type of apparatus is given in the following figure.

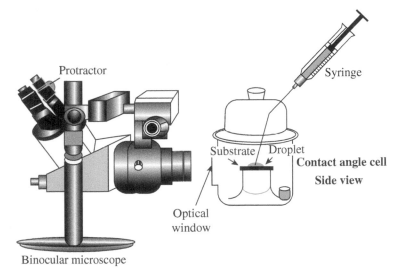

Figure 2.30 Schematic diagram of the contact angle apparatus.

The glass cell and syringe should be cleaned by ethanol rinsing, followed by drying using nitrogen. The glass plate (which must be cleaned in the manner described later) is then positioned on the upraised table, and a small beaker of the liquid to be used is placed in the base of the cell. The cell is then sealed and left to equilibrate for 5 mins. About 0.5 to 1.0 cm^3 of the liquid is filled into the syringe (using care to remove all air bubbles), and the steel needle is lowered to within less than 1 mm of roughly the centre of the plate (use the microscope for this). A droplet of the liquid is then slowly forced out and allowed to equilibrate. Since the needle is left in the droplet during measurement, the droplet must be of sufficient size so that the region near the TPL line is not affected by the needle changing the shape of the droplet. The drop volume should then be slowly increased until the maximum angle is obtained just before the TPL moves forwards. (This has to be repeated several times to obtain the maximum angle). Measure the contact angle with the TPL at different positions on the plate. Take the average value to give the advancing contact angle (θ_A).

Follow a similar procedure to measure the receding angle (θ_R), but slowly withdraw liquid into the syringe and measure the minimum angle just before the TPL moves.

Sample preparation

Before methylation, the soda glass plates must be cleaned by washing with warm 10% NaOH solution and then rinsed in high-quality distilled water, finally blow-drying in nitrogen. Be careful and always wear safety glasses. Warm concentrated NaOH solution is *very harmful* to the eyes. The contact angle of water on *clean* glass should be very low; if not, then further cleaning is required. The cleaned glass is very easily contaminated by finger grease and exposure to laboratory air. For these reasons, the samples have to be prepared just before they are used in the cell and must only be handled using clean tweezers.

Methylated (hydrophobic) glass is prepared simply by exposing a clean, dry plate to the vapour of highly reactive trimethylchlorosilane (Me$_3$SiCl) for about one minute in a fume cupboard. Simply place the plate with the polished surface exposed in a large clean beaker containing a smaller beaker of liquid Me$_3$SiCl. Loosely cover the large beaker. The Me$_3$SiCl reacts vigorously with water as well as the surface silanol groups on glass:

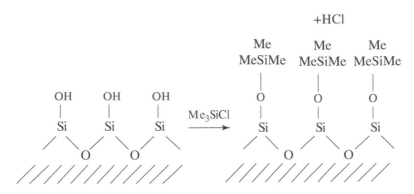

Figure 2.31   Methylation of the silica surface.

The hydrochloric acid vapour evolved is toxic, and all reactions should be carried out in a fume cupboard.

The resulting monolayer of methyl groups is chemically attached and completely alters the wetting properties of the surface. Again, to prevent contamination these plates must be handled only with tweezers and stored in cleaned, sealed containers.

Liquids used to determine the critical surface tension of methylated glass

Measure advancing and receding contact angles on methylated glass plates using the liquids in the order given in the following table.

Liquid	γ_{LV} /mJm^{-2}
water	73
formamide	56
methylene iodide	49
ethylene glycol	47.5
propylene carbonate	41
di-methyl aniline	36.5

For all these liquids except water, prepare the syringe and cell in a fume cupboard and seal before measuring contact angles in the laboratory (use caution). Use the same methylated plate for the first two liquids and a second plate for the other three. In each case, rinse the plate with clean ethanol and blow dry when changing liquids.

Plot out $cos\theta_A$, $cos\theta_R$ and $cos\theta_{AV}$ against the corresponding γ_{LV} values to estimate the value of γ_c.

At the end of the experiment, clean the cell and syringe thoroughly using ethanol, and blow dry with nitrogen.

FOR CONSIDERATION/TYPICAL QUESTIONS

(1) Why is it that only a very thin monolayer (~5 Å) of adsorbed methyl groups completely alters the macroscopic water wetting properties of glass?

(2) What do you think are the causes of the contact angle hysteresis observed in this experiment?

(3) What would you expect for the wetting properties of these liquids on untreated clean glass?

(4) Is the value you obtained for γ_c reasonable for this type of surface?

3

The Prevention of Fluid Cavitation

Cavitation is the main cause for the limitation of speed in fast ocean-going vessels due to cavitation effects on propellers. The main approach to solving this issue is to select operating conditions, geometric design and wear-resistant materials to mitigate against these detrimental effects. However, dissolved atmospheric gases facilitate the nucleation of cavities in fluids, and it has been recently demonstrated that removal of these dissolved gases dramatically reduces cavitation.

A SHORT HISTORY OF CAVITATION IN FLUIDS

At the end of the 19th century, it was realised that ships were not attaining their design speed, and the reason was later found to be cavitation. The effects of the collapse of a spherical cavity within a fluid were first considered by Besant in 1859, and cavitation in fluids has been studied for over a century since this pioneering work and that of Reynolds in 1886 and Lord Rayleigh in 1917. There are two main types: inertial cavitation, created by fluid flow in pumps, valves and propellers; and non-inertial cavitation, created by oscillatory processes, such as sonication. Cavitation also occurs in fine cavities between solid surfaces.

In many diverse processes, cavitation not only reduces fluid flow efficiency, but the collapse of the bubbles created near surfaces creates shock waves which cause cavitation damage or wear. The temperatures produced by the rapid collapse of a bubble can reach 20,000 K and can cause transient light emission or sonoluminescence. Fast cavity formation produces partially

Applied Colloid and Surface Chemistry, Second Edition. Richard M. Pashley
and Marilyn E. Karaman.
© 2021 John Wiley & Sons Ltd. Published 2021 by John Wiley & Sons Ltd.
Companion website: www.wiley.com/go/pashley/appliedcolloid2e

filled cavities with low internal pressure, which can then explosively collapse when the reduced fluid pressure returns to its (higher) static value.
Slow cavity formation allows dissolved gases to diffuse into the cavity to
produce stable bubbles. Boiling is defined as the formation of cavities within
a fluid as the temperature is increased at constant (atmospheric) pressure,
whereas cavitation occurs when the fluid pressure is reduced (locally) at a
constant temperature. Inception is when the cavitation effect just begins.

The cavitation index C (σ or sometimes κ) is used as a measure of cavitation potential and is defined as:

$$C \equiv \frac{(p_r - p_v)}{\frac{1}{2}\rho V^2},\qquad(3.1)$$

where $p_r - p_v$ is the pressure difference due to dynamic effects of the
fluid flow, p_r is the local (reduced pressure), p_v is the vapour pressure of
the fluid (at that temperature), ρ is the density of the fluid, and V the fluid
velocity. In effect, the cavitation index is the ratio of the work done by the
pressure differential to the work corresponding to the fluid flow kinetic
energy. The parameter has a critical value denoted as C_i, which is when cavitation inception occurs, and hence when $C < C_i$ this corresponds to conditions of advanced cavitation. In general, when $C \gg 1$ cavitation is
increasingly unlikely to occur.

In industry, the effects of cavitation in fluid systems, cavitation wear, fluid
oxidation/degradation and fouling in aqueous systems are dealt with via
design compromises and the development of new materials, coatings, various fluid additives and regular maintenance. It is well known that cavitation
phenomena have an effect on the efficiency of fluid transmission and lead to
material degradation in hydraulic systems. All fluid pumping installations
are designed to prevent cavitation in order to ensure steady fluid flow
through piping. Furthermore, the maximum shaft speed of centrifugal
pumps and therefore the fluid velocity, flow rate and head obtainable from
them are limited by the need to ensure that cavitation does not occur at the
leading edge of the pump impeller(s). Otherwise, unsteady and variable
mass flow, noise, vibration and, in some instances, 'choking' or flow collapse
results once sufficient air bubbles form at the impeller hub under the action
of centrifugal forces.

In fluid cavitation, it is generally assumed that local suction pressures just
below the vapour pressure of a fluid, at a given temperature, will nucleate
bubbles which then implode once local hydrostatic pressure returns. In
practice, there is a significant additional barrier to the formation of cavities

in the absence of suitable nucleation sites, and it is actually very difficult to cavitate pure liquids in clean, smooth vessels. Nano-sized cavities are usu-ally the smallest structures which can be considered as a separate phase, and their growth or collapse controls the extent of cavitation. The presence of dissolved atmospheric gases facilitates fluid cavitation. For example, water can dissolve close to 20 mL of atmospheric gases per litre, and oils typically ten times more. The removal of these dissolved gases inhibits fluid cavita-tion. The prevention of fluid cavitation by degassing can improve a wide range of processes, such as pumping efficiency, reduce corrosion and wear in fluid systems, prevent oxidation and degradation of liquids and assist in the prevention of inorganic fouling related to corrosion. Degassing will also reduce bio-fouling and related microbial growth in liquids and gels.

The importance of dissolved atmospheric gas molecules as nucleation sites for cavitation can be readily demonstrated by studying the properties of degassed fluids. Even using simple theory, we can expect this behaviour in advance. Consider the intermolecular forces which must operate between water molecules to hold them together – to make water the most important liquid on earth. To do this we can consider the pressure required to force water molecules into the liquid form simply by using the ideal gas equation:

$$PV = nRT \qquad (3.2)$$

or

$$P = \rho RT \qquad (3.3)$$

where ρ is the density of liquid water at say 20 °C. The estimated tensile pressure holding liquid water together is therefore about 1350 atm! However, we know from simple laboratory experiments that water bubbles and cavitates when we apply a suction pressure of only one atmosphere. This enormous difference must be due the presence of dissolved atmospheric gas molecules, and we will see further direct evidence for this later.

In 1982, Israelachvili and Pashley made the remarkable discovery that hydrophobic surfaces immersed in water were attracted to each other with a long-range (~10 nm) force much stronger than expected for van der Waals forces. They called this the 'long-range hydrophobic interaction'. Since that discovery, there have been many reports extending the range of the force, depending on surfaces and conditions, in some cases to a range of 300 nm. These observations created a problem for theoretical interpretation simply because the local effect of a hydrophobic, non-hydrogen bonding surface on water should only extend a few water molecules, at most. In 1985, Pashley et al. noticed that a bridging cavity was formed when two solid hydrophobic surfaces were pulled apart in water, and later Christenson et al. reported cavitation as two hydrophobic surfaces approached within a few nm, but before making contact. These observations led to the suggestion that the long-range hydrophobic interaction may be caused by the formation of bridging cavities between hydrophobic surfaces, giving a more reasonable explanation for the extraordinarily long range of the force.

If cavitation held hydrophobic surfaces together, then it follows that the inhibition of cavitation by removing dissolved atmospheric gases may indeed allow oil and water to mix. This idea was first tested by us in experiments carried out in 1996 by studying the effects of degassing on the dispersion of oil droplets in water. Hydrocarbon oils such as decane immediately phase separated on shaking with water, but on degassing, a fine stable dispersion was easily produced. This work demonstrated that cavitation also plays a central role in the dispersion of oil in water. The presence of dissolved non-polar gas molecules in the surrounding fluid, in equilibrium with the atmosphere, must produce nucleation sites throughout this fluid. Both aqueous and non-aqueous fluids show similar effects, and in addition, non-polar fluids have an increased capacity to dissolve gases relative to water. It was demonstrated in this work that degassing prevents cavitation.

The modest level of degassing required can be achieved using commercial membrane processes and other methods. A fluid degassing pre-treatment process could therefore improve the efficiency of a wide variety of technologies that employ fluids and, as a secondary benefit, eliminate wear caused by cavitation. The application of this technology could be highly beneficial to the efficiency of propellers. For example, propeller cavitation in ships and submarines is typically controlled by reducing rotation rate and/or blade pitch, whereas we demonstrate here that cavitation can be completely prevented by releasing degassed water adjacent to the low-pressure side of a rotating propeller, without varying blade speed or pitch.

THEORETICAL PREDICTION OF THE LINK BETWEEN DEGASSING AND CAVITATION PRESSURE

In fluid cavitation, it is generally assumed that local suction pressures just below the vapour pressure of the fluid, at a given temperature, will nucleate bubbles which then implode once local hydrostatic pressure returns. In practice, there is a significant additional barrier to the formation of cavities in the absence of suitable nucleation sites, and it is actually very difficult to cavitate pure liquids in clean, smooth vessels. Nano-sized cavities are usually the smallest structures which can be considered as a separate phase, and their growth or collapse controls the extent of cavitation. Both homogeneous cavitation and heterogeneous nucleation in water are considered here, caused by the presence of inert dissolved atmospheric gas molecules throughout the bulk liquid phase.

The barrier to 'ideal' homogeneous cavitation can be estimated from a simple analysis of the formation of a nano-sized spherical cavity. The total energy E_T of a cavity of radius r is given by the sum of the negative work done by the suction pressure $-\Delta P$ on the cavity volume and the surface tension work done on creating the surface of the cavity. Thus, the total cavity energy is given by:

$$E_T = \frac{4}{3}\pi r^3 (\Delta P) + 4\pi r^2 \gamma \qquad (3.4)$$

A diagram of the behaviour expected for water is shown below. Assuming that nm-sized cavities must form within pure water, the barrier to their formation is very high, of the order of 100 kT, which makes their formation difficult.

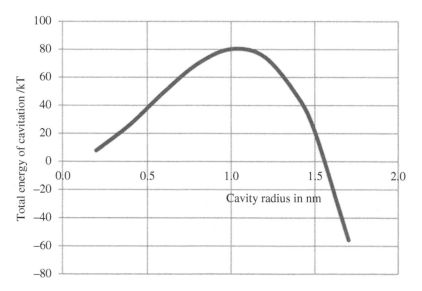

Figure 3.1   This diagram illustrates the theoretical calculation of the energy (in kT units) required to form a spherical cavity of radius r in pure water under ideal, degassed conditions, in the absence of nucleation sites, with an applied suction pressure of −1400 atm.

If we make the assumption that 1 nm is the critical radius of cavity formation, i.e. when $dE_T/dr = 0$, we can estimate the critical suction pressure from
the Laplace equation:

$$\Delta P = -\frac{2\gamma}{r_c}. \tag{3.5}$$

This gives a critical suction pressure of about $-1,460$ atm for pure water
or -500 atm for a typical hydrocarbon liquid. The largest suction pressure
observed experimentally for degassed water was -1400 atm, which is quite
close to the theoretical prediction obtained simply from using the ideal gas
equation given earlier.

In most practical situations, contaminants and 'real', rough surfaces facilitate the heterogeneous nucleation of cavities in water at much lower suction
pressures than this. The presence of dissolved gases and hydrophobic groups
also substantially reduces the cavitation pressure. Experimental cavitation
pressures are typically about -1 atm for distilled water saturated with air,
since this is effectively the vapour pressure of water at room temperature.
However, a suction pressure of -200 atm is required when the water is
99.98% degassed according to Figure 3.2. From these experimental results,
it is clear that degassing water strongly inhibits cavitation, especially when
degassed to greater than 99%. The disruptive presence of dissolved non-
polar gas molecules in the bulk liquid water phase apparently produces
nucleation sites throughout the liquid. An equivalent result occurs with non-
aqueous fluids, which also have an increased capacity to dissolve gases relative to water.

A theoretical model can be developed to estimate the cavitation pressure
p_c required to cause (heterogeneous) cavitation in water at a wide range of
dissolved gas levels. The basic principle used is that the pressure required
can be estimated from the change in activation energy $\Delta\mu$ required to transfer a dissolved gas molecule (e.g. N_2) from the aqueous phase to the gas
phase. The chemical potential of dissolved gas in water, $\mu(g,w)$, will change
with the concentration of the dissolved gas in water, x_g, (i.e. mole fraction),
and this is given by:

$$\mu(g,w) = \mu^0(g,w) + kTlnx_g, \tag{3.6}$$

and since the gas-phase chemical potential will stay constant, the activation
energy ($\Delta\mu$) for the transfer of the dissolved gas from solution to gas phase
will also vary with concentration in the aqueous solution as:

$$\Delta\mu = kTln\left(\frac{x_g^s}{x_g}\right), \tag{3.7}$$

where x_g^s is the mole fraction of gas in the water under standard atmospheric conditions (i.e. 1 atm).

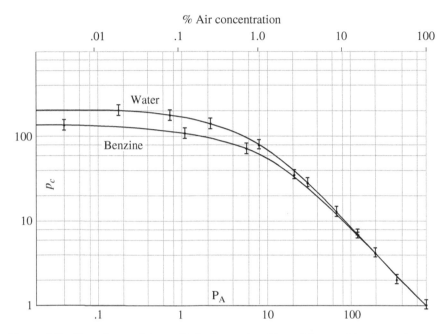

Figure 3.2 Experimental data which shows the effect of degassed levels on the suction pressures required to produce cavitation in water and benzene. It is reproduced from Galloway, W. J., 'An Experimental Study of Acoustically Induced Cavitation in Liquids.' *The Journal of the Acoustical Society of America*, 1954. 26(5): 849–857.

Since the cavitation pressure can be estimated from the activation-energy equation:

$$p_c = p_0 \exp\left(\frac{\Delta\mu}{kT}\right), \tag{3.8}$$

where p_0 is the standard, i.e. saturated-gas cavitation value, of 1 atm, it follows that:

$$p_c = p_0\left(\frac{x_g^s}{x_g}\right). \tag{3.9}$$

Note that for air-equilibrated water, x_g^s is about 1.53×10^{-5} and also note that this result reduces simply to:

$$p_c\left(atm\right)=\left(\frac{100}{100-\%deG}\right). \qquad (3.10)$$

This cavitation-pressure model will work for any liquid, since it is based entirely on gas solubility relative to a standard state. Graphical representations of this result are given in the following figures.

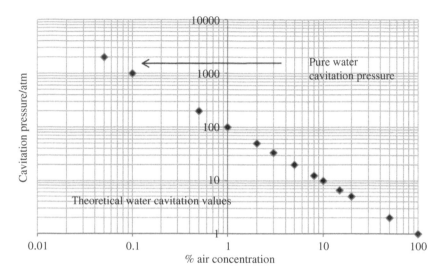

Figure 3.3   Calculated cavitation pressures for water obtained using Equation (3.10).

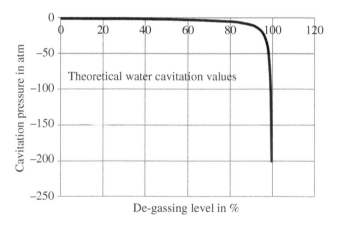

Figure 3.4   Effect of degassing on cavitation pressure in water.

For the case of water, the model predicts that, when the dissolved gas level is reduced to about 0.07% (or about 0.6 µM), the gas cavitation pressure equals that of pure water. (Note that at 20 °C, water in equilibrium with the atmosphere has 0.85 mM of dissolved gases: N_2, O_2, CO_2 and Ar.) The pure water cavitation pressure can be estimated using the Laplace equation with an estimated, critical cavity radius of 1 nm:

$$\Delta P = -\frac{2\gamma}{r}. \tag{3.11}$$

This gives a calculated (homogeneous) cavitation pressure of 1460 atm. This value agrees with the reported experimental value for pure (i.e. completely degassed) water of –1400 atm. These high suction pressures are also consistent with the basic (kinetic) model of condensed liquids, in which the repulsive ideal gas pressure ($P = nRT/V$) is more than balanced by the cohesive molecular forces. For water at 20 °C, the 'ideal' repulsive molecular pressure $P = 1353$ atm. Cohesive pressures in liquid water must be higher than this. Thus, there is strong evidence for this high value for pure, completely degassed water.

This analysis strongly suggests that the Galloway data (1954), shown earlier, is correct at relatively low degassed levels (less than 99%) but that the cavitation pressures expected, for pure liquids, at higher levels of degassing were most likely not achieved in these experiments, probably due to the ubiquitous presence of contamination particles, which offer nucleation sites. Thus, the reported maximum of about –200 atm is much less than the theoretical prediction.

Experimental study of the prevention of propeller cavitation in degassed water

The cavitation created by a simple three-blade propeller with a maximum angular speed of 2960 rpm can be observed in a water-filled perspex rectangular tube, as shown in the following photos. Tap water was pumped at 600 mL/min to the membrane with a vacuum at the outer chamber of the membrane unit. The transmitter connected to the oxygen electrode showed that the oxygen content in the water tank continuously decreased.

Figure 3.5 Photograph of a propeller used to study cavitation in the laboratory.

Figure 3.6 Cavitation occurring in air-equilibrated water at atmospheric pressure at a propeller rotation rate of 2960 rpm.

Visible cavitation occurred in air-equilibrated water at atmospheric pressure with a propeller rotation rate of 2960 rpm. This is similar to the effects observed when seawater is subjected to compressive pressures in the 10–55 atm range. On release of this pressure through a needle valve, cavitation of

Figure 3.7 Cavitation produced in the left-hand beaker following pressure release of normal atmospheric-equilibrated seawater. The right-side beaker was produced using 99.5% degassed seawater.

dissolved gases occurred, as illustrated in the photo above. The left-hand beaker contains atmospheric-equilibrated seawater collected following high-pressure valve release, and this is compared with 99.5% degassed seawater on the right side. Clearly, degassing the seawater prior to pressure release completely prevents cavitation.

These initial studies have recently been extended and applied to the cavitation formed by a rapidly rotating propeller. In these studies, it was found that, on increasing the degassed levels of the surrounding water, the speed of a propeller could also be increased before cavitation was observed, as illustrated in the following photo. Typical results of these rotation-rate/degassing measurements are summarised in Table 3.1, which clearly shows that higher degassing levels produced higher cavitation pressures in the system (these data were obtained from the theoretical values for water cavitation discussed earlier). The results given in Table 3.1 were also reproduced using a 0.5 M NaCl solution to simulate seawater. As an example, 70% degassing was found to be sufficient to prevent cavitation even at the maximum rotation rate of 2960 rpm, as shown in the following photo. This degree of degassing corresponds to a cavitation pressure of about 3 atm, according to Eq. (3.10).

Figure 3.8 Complete cavitation prevention after gassed water was replaced with 70% degassed tap water (at 2960 rpm propeller).

Table 3.1 Experimental results of the degassing effects on cavitation observed for a rotating propeller completely immersed in degassed water, at different degassing levels.

Time (min)	Degassing (%)	Cavitation pressure (atm)	Minimum rpm to start cavitation
0	0	1	–
5	19.7	1.3	148
10	27.8	1.5	592
15	36.9	1.6	1628
20	45.4	1.8	2072
30	56.0	2.8	2500
60	76.7	4	Not even with 2960
90	84.4	7	Not even with 2960
120	87.0	8	Not even with 2960

In other experiments, degassed water was introduced locally, close to the low-pressure side of rotating propeller blades immersed in gas-equilibrated water. Inside the perspex tube, gassed tap water flowed freely through a bent metallic pipe, and degassed (80%) water (at a flow rate of about 1/10th the

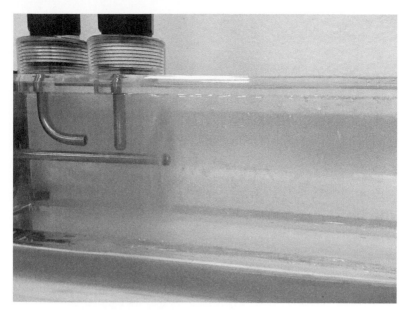

Figure 3.9 Cavitation occurring in flowing tap water at atmospheric pressure, i.e. in fully gassed water (at 2300 rpm).

background water flow rate) was released through a vertical metallic pipe with suitable holes, as shown in the following photo. With gassed tap water flowing within the perspex tube, visible cavitation occurred, as can be seen in the following photo.

However, when a local flow of degassed water was released adjacent to the propeller blades, all cavitation ceased, as shown in the following photo. These experiments were carried out at high rotation rates of 2300 rpm using a three-blade propeller. The sound associated with cavitation also ceased on starting the degassed water flow. Audible measurement results showed that the cavitation noise was reduced from 75 dB to 65 dB after releasing the degassed water behind the propeller, despite having another additional noise involved, which was from the water pump releasing the degassed water continuously. The observed drop in dB level, even with the additional pump noise, corresponds to a sound intensity decrease, on releasing the degassed fluid, of 10x. Similar results were also obtained using a 0.5 M NaCl aqueous solution to simulate seawater.

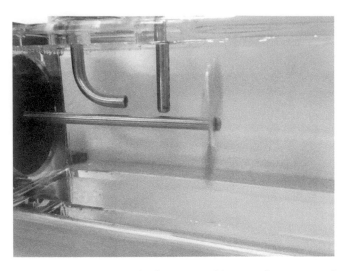

Figure 3.10 Cavitation was completely prevented in gassed tap water after flowing 80% degassed water directly onto the rotating propeller blades (at 2300 rpm).

Effects of degassed boundary-layer films on a rotating propeller blade

When the flow of degassed water was terminated in these experiments, it was observed that the cavitation effects remained absent for several seconds or about 100 to 200 rotations of the propeller. This is consistent with the observation (shown in the photo) that release of degassed water onto the upper rotating propeller blade prevented cavitation effects through a full cycle of the blade. These observations support the view that coating a rotating propeller with a film of degassed fluid is sufficient to give protection against cavitation.

The lowest pressure regions on the low pressure side of a rotating propeller blade are at the following edge and the tip of the blade. For zero-slip conditions, water flow across a plate of length x (i.e. a propeller blade), water flow will slow down close to the surface; this effect will increase with an increase in x, as shown in the following diagram.

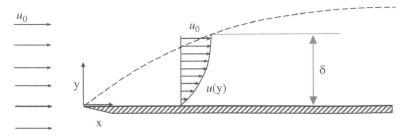

Figure 3.11 Schematic diagram of stationary or boundary layer formation as a fluid flows over a flat solid surface.

The experimental boundary layer equation can be used to calculate the thickness δ of the water layer next to the plate at which the velocity reaches 99% of the bulk water flow velocity:

$$\delta = 5x / \sqrt{R_e}, \qquad (3.12)$$

where R_e is the Reynolds number. This situation is shown in the diagram, illustrating the 'zero-slip' condition, in which the fluid flow slows down in proximity to the surface; the speed of the fluid depends on the distance x the fluid has moved over the surface and the thickness δ of the fluid layer.

The Reynolds number in this case is given by

$$R_e = (\rho V_\infty x) / \mu, \qquad (3.13)$$

where ρ is the density of water, V the bulk fluid velocity (i.e. at infinite distance from the static plate), μ is the dynamic viscosity of water and x is the length of the surface.

Typically, R_e will be high, corresponding to laminar flow, in the range 1000 to 5×10^5. For example, if R_e = 50,000, the boundary layer water thickness δ will be about 1.1 cm for an x value of 0.5 m. Much closer to the surface, say within 1% of the boundary layer thickness, the water moves only slowly relative to the solid surface: for example, in a water layer of 110 μm thickness, the water flow velocity is about 10 cm/sec.

Cavitation occurs when dissolved gas is present in a fluid such as water. For the diffusion of gas in one direction (x), the appropriate equation is Fick's second law in the form:

$$\frac{\partial C}{\partial t} = D \frac{\partial^2 C}{\partial x^2}, \qquad (3.14)$$

where C is the solute concentration and D the diffusion coefficient. When completely degassed water is exposed to air at atmospheric pressure, a very thin layer will form rapidly at the surface which will be at the atmospheric dissolved gas concentration (of about 1 mM). This layer will maintain its saturated concentration from then onwards (in equilibrium with the atmosphere), and the dissolved gas will diffuse further into the water. Thus, Fick's law must be solved for the boundary conditions $C = C_0$, when $x = 0$ for any value of t and $C_x = 0$ for $x > 0$ when $t = 0$. Also, $C_x = C_0$ for any value of x as t becomes very large. The solution to Fick's law under these conditions is:

$$C = C_0 \left[1 - \mathrm{erf} \left(\frac{x}{2\sqrt{Dt}} \right) \right] \qquad (3.15)$$

where the error function can be calculated using either tables or the series:

$$\mathrm{erfx} = \frac{2}{\sqrt{\pi}} \int_0^x e^{-t^2}\, dt = \frac{2x}{\sqrt{\pi}}\left[1 - \frac{x^2}{1\times 3} + \frac{x^4}{2\times 1 \times 5} - \frac{x^6}{3\times 2 \times 1 \times 7} + \ldots\right]. \quad (3.16)$$

For oxygen and nitrogen gases in water, the value of the diffusion coefficient D at 20 °C is about 2×10^{-5} cm²/s and the saturated gas concentration, C_0 is about 1 mM. Using these values, the results shown in the following graph were obtained, which gives the degassed level as a function of diffusion relative to film thickness. These calculated results indicate that, for quiescent water, that is in the almost stationary part of the boundary layer, closest to the solid surface, significant re-gassing of a degassed water film of about 100 μm thickness would take several seconds.

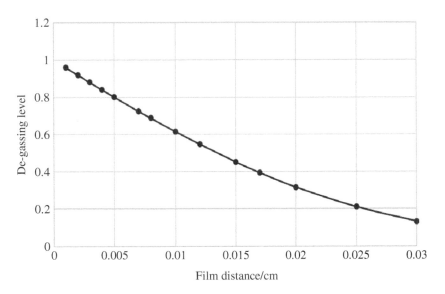

Figure 3.12 Calculated degassing level produced in a thin (quiescent) film of initially degassed water when re-exposed to the atmosphere for several seconds.

Thus, a transient coating of a thin boundary layer of degassed water on a rapidly rotating blade could be expected to prevent cavitation effects for many subsequent rotations, as has been observed experimentally. This means that the amount of degassed water required to prevent or minimise cavitation can be substantially reduced, and the effect of cavitation minimisation can persist for a significant period of time, e.g. a few or more seconds. The effect of reducing the occurrence of cavitation should persist until the stationary film drains as the blade rotates or until diffusion from the bulk fluid re-gasses the degassed film.

The effects of having a transient film of degassed fluid could be further opti-mised by having a periodic controlled release of a positive, relatively high-pressure flow of degassed fluid, released close to the rotating blade and timed to strike close to the leading edge on the negative-pressure side. Once the degassed fluid strikes the blade, it would coat the blade (transiently) with a thin film of degassed fluid (e.g. water) to form the boundary layer. The zero-slip boundary condition expected for most blade surfaces will ensure the transient retention of a thin, degassed, film next to the rotating surface. Put another way, the degassed fluid could be periodically directed to the surface using a pulsed flow to form the boundary layer. Use of a pulsed flow of degassed fluid may mean degassed fluid only needs to be released in a periodic fashion, such as every 1 in 100 rotations of the blade, or for only 1% of the time.

The leading edge of the rotating blade generates a high positive pressure which suppresses cavitation. As the blade rotates, the thin stationary film of degassed water tends to drain towards the following (low-pressure) edge of the blade due to the shear forces of the water flowing over the blade, where it continues to suppress cavitation. At the same time, dis-solved gases within the flowing fluid (e.g. water) begin to diffuse into the thin boundary layer of degassed water. However, gas diffusion under effec-tively quiescent conditions next to the solid surface is relatively slow; for example, for a 0.1 mm film, this re-gassing will take several seconds.

In addition to these effects, the centrifugal forces generated by the rotat-ing blade will also tend to force the stationary initially degassed boundary layer to flow towards the tip of the rotating blade, which will also help to prevent cavitation at this point, where it is often observed.

The zero-slip condition expected for most propeller materials can be ensured by selection of a suitable hydrophilic surface capable of surface bind-ing to water. This will aid the formation of a boundary layer. Hydrophobic materials, such as Teflon, can produce slip conditions which are not conducive to the formation of boundary layers; hydrophobic surfaces are well-known to aid cavitation. Micro-surface roughness could also be used to enhance bound-ary-layer formation by reducing surface flow across the propeller.

Proposed new cavitation number to include the effects of fluid degassing

These theoretical and experimental results lead to the obvious suggestion that an extended cavitation number could be developed to include the effects of fluid degassing, which acts to oppose the effects of local negative pressure on cavitation formation. The primary effect of degassing is to increase the cavitation pressure p_c, and this factor can be easily included in a new defini-tion of an extended cavitation number, C^{dG}, where:

$$C^{dG} \underset{=}{\mathrm{def}} \frac{p_c + p_r - p_v}{\frac{1}{2}\rho V^2},$$

(3.17)

which, from Eq. (3.9), becomes:

$$C^{dG} = \frac{p_0\left(\dfrac{x_g^s}{x_g}\right) - p_0 + p_r - p_v}{\frac{1}{2}\rho V^2}.$$

(3.18)

This new definition incorporates the effects of degassing, equally weighted with the opposing effects of reduced fluid suction pressure, on net cavitation. When the fluid is completely gassed, this relation reduces to the standard version, i.e. Eq. (3.1). Typical results are given in the following graph for the case in which degassing can dominate the cavitation number against the cavitation effect of a local pressure, which was reduced significantly below the vapour pressure. That is, this graph shows the effects of degassed level on the cavitation number for water of flow velocity 20 m/s, with normal water vapour pressure (0.03 atm at room temperature) and a reference local pressure (p_r) of 0.01 atm. Under these conditions, cavitation becomes increasingly unlikely as $C^{dG} \gg 1$, that is, at degassed rates above about 80% in this case.

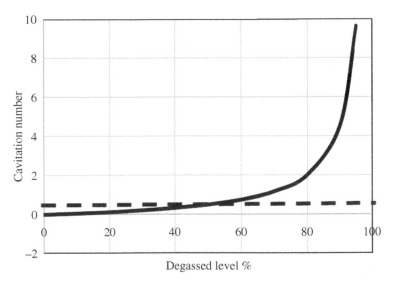

Figure 3.13 Proposed cavitation number applied to systems with different degassing levels.

This new cavitation number could be used to design conditions to reduce noise generation and wear in many fluid flow situations by using degassed fluid to inhibit the cavitation process for pumps and propellers in ships and submarines.

Industrial Report

Ship propeller cavitation

Ship and boat propeller cavitation is the formation of bubbles of dissolved gas released into the water by the low or negative pressure generated on the front of propeller blades as they deliver propulsion thrust causing high pressure on the back of the blades. These bubbles last for short durations in the turbulent flow past the propeller and they generate and collapse causing hammer-like impact loads on the propeller blades and rudders often in the order of 7 kg/ cm^2. These impacts cause damage to the propeller blade and rudder surfaces, referred to as erosion.

The severe effects of cavitation and bubble collapse include noise across a range of low frequencies propagated for long distances in the water, vibrations mainly in the rear sections of the ship and material damage to propellers and rudders. High noise and vibrations levels and intensities affect crew and passenger comfort and stability; cavitation erosion could lead to a significant reduction of propeller efficiency and structural integrity. Loss of propeller efficiency will lead to loss of the fuel efficiency of propulsion and impact the economics of merchant shipping.

Marine traffic in the world's oceans is increasing. This includes marine craft ranging from small boats to large ships. Merchant ships are increasing in number as well as size, linked to overall economic growth. In the 60 years from 1948 to 2008, the number of ships travelling the oceans and waterways increased by 250%, gross shipping tonnage by 900% and, in the 20 years from 1992 to 2012, shipping density by 300%. Ship-generated noise has increased in proportion to these trends, becoming the most ubiquitous and pervasive source of human-generated noise in the oceans. It is likely that boat and ship noise is massively impacting a wide range of marine mammal species, and these impacts continue to be of great concern.

Propeller cavitation noise is a great challenge to naval shipping of all classes including warships, aircraft carriers and submarines. Quiet navigation and stealth is a highly priced attribute for all such craft. If propeller cavitation could be eliminated, it would also allow the design of naval and merchant ships with smaller propellers and higher speeds resulting in improved ship design and great strategic and economic advantage.

The shipping industry, since the beginning of motorised shipping, has placed great importance on finding ways to reduce, if not eliminate, propeller-generated cavitation. These efforts continue to the present time. Vast research and development effort has been directed to improving the cavitation reduction performance of ship and boat propellers.

Many innovations have been developed and applied to reduce cavitation and its impacts. These are focussed on strategies such as blade geometry modification, propeller hub modification and wake inflow modification. Computational fluid dynamics–based modelling is a frequently used tool in these designs and innovations. A very recent such innovation is to drill holes on the propeller blades at optimal locations determined by CFD modelling.

Despite all of the R&D particularly in the past 100 years, a complete and satisfactory solution to ship propeller cavitation and its impacts has remained beyond the grasp of the shipping industry and the related R&D establishments of the world. Finding such a solution remains of paramount importance for the shipping industry, defence industry, world trade and commerce, the environment and all marine life.

Mr Satis Arnold
BSc Eng Peradeniya, MEngSc UNSW
Director, GreenTech Human Resources
Development Consultants Pvt Ltd
Director, Cambridge Partners Pty Ltd
55 Gellibrand Street
Campbell ACT 2612
Australia

Experiment 3.1

Membrane degassing for water and aqueous solutions

Hollow-fibre membranes can be purchased from Membrana, Charlotte, USA (Model 2×6 Radial Flow Superphobic). A vacuum pump, Fossa FO 0015 A (Busch Sydney, Australia) can be used for degassing. A diaphragm water pump (model: FloJet-D3732-E5011) can be purchased from CreativePumps Australia. An InPro 6900 oxygen electrode system can be obtained from Mettler-Toledo Ltd., Melbourne, Australia.

Hollow-fibre hydrophobic membrane systems offer the most efficient commercial process for degassing. A photograph of the experimental setup used in this study is given in Figure 1. A high vacuum is applied to the outside of a hollow-fibre hydrophobic membrane or membrane array where the dissolved gases emerge, while the aqueous solution flows through the core of the hollow fibres. The hollow-fibre membranes are strongly hydrophobic (Teflon or polypropylene), and have small pores designed to prevent liquid water passing through them due to the Laplace pressure. Only water vapour exists in the pores. The high-surface-area membrane efficiently transfers water vapour and atmospheric gases. This technique has been used to produce a continuous flow of water more than 99.5% degassed. The vacuum pump, with ultimate pressure $\leq 2.5 \times 10^{-2}$ mbar, should be protected from exposure to water vapour using two 5 L Pyrex glass tanks containing pre-dried granular silica gel. The dissolved oxygen level in the tap water, measured using the InPro 6900 oxygen electrode system with a detection limit of 1 ppb, can be used to monitor the degree of water degassing in these experiments.

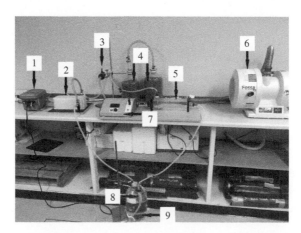

Figure 3.14   Photograph of a system used to study membrane degassing effects on cavitation: (1) M700 process analysis system, (2) hollow-fibre membrane, (3) water pump, (4) silica gel tanks, (5) observation cell, (6) vacuum pump, (7) variable motor, (8) DO electrode, and (9) degassed water reservoir.

Experimental evaluation of degassing in the prevention of cavitation

The properties of degassed water produced using this membrane system can be compared with normal atmospheric-gas-equilibrated water by observation of the effect of reducing pressure, using either a water jet pump or a 1 atmosphere mechanical pump, connected to a flask with a piece of clean Teflon tubing immersed in a water-filled Pyrex Buchner filter flask, as shown in Figure 2. Note that using a basic mechanical pump will require some protection from water vapour using a silica gel absorbent, as shown in Figure 2.

Figure 3.15 Vacuum pump system to study the effects of degassing on cavitation in water.

Make observations of the difference between bubble formation at the surface of the Teflon when immersed in atmosphere-equilibrated distilled water and seawater (i.e. 0.5 M NaCl solution) and then with degassed water, that is, as the applied reduced pressure approaches the vapour pressure of water, i.e. at a pressure of about 20 mm of Hg or 20/760 or about 3% of an atm.

QUESTIONS

1) What is the lowest pressure attainable by a water jet pump?
2) At what reduced pressure would you expect to see bubble formation?
3) What do you observe without the Teflon tubing? Why is this different?
4) Could you estimate the degassing level required to produce spontaneous bubbling on other surfaces?
5) Was the observed bubbling different in distilled water and saltwater, and if so, why?

4

Thermodynamics of Adsorption

Derivation of the Gibbs adsorption isotherm. Determination of the adsorption of surfactants at liquid interfaces. Laboratory project to determine the surface area of the common adsorbent powdered activated charcoal.

BASIC SURFACE THERMODYNAMICS

For an open system of variable surface area, the Gibbs free energy must depend on composition, temperature, T, pressure, p, and the total surface area, A:

$$G = G\left(T, p, A, n_1, n_2, \ldots\ldots n_k\right) \qquad (4.1)$$

From this function it follows that:

$$dG = \left(\frac{\partial G}{\partial T}\right)_{p,n_i} dT + \left(\frac{\partial G}{\partial p}\right)_{T,n_i} dp + \left(\frac{\partial G}{\partial A}\right)_{T,p,n_i} dA + \sum_{i=1}^{i=k}\left(\frac{\partial G}{\partial n_i}\right)_{T,p,n_i} dn_i \quad (4.2)$$

The first two partial differentials refer to constant composition, so we may use the general definitions

$$G = H - TS = U + PV - TS \qquad (4.3)$$

Applied Colloid and Surface Chemistry, Second Edition. Richard M. Pashley and Marilyn E. Karaman.

to obtain

$$\left(\frac{\partial G}{\partial T}\right)_{p,n_i} = -S \qquad (4.4)$$

and

$$\left(\frac{\partial G}{\partial p}\right)_{T,n_i} = V \qquad (4.5)$$

Insertion of these relations into equation (4.2) gives us the fundamental result:

$$dG = -SdT + Vdp + \gamma dA + \sum_{i=l}^{i=k} \mu_i dn_i \qquad (4.6)$$

where the chemical potential μ_i is defined as:

$$\mu_i \equiv \left(\frac{\partial G}{\partial n_i}\right)_{T,p,n_j} \qquad (4.7)$$

and the surface energy γ as:

$$\gamma \equiv \left(\frac{\partial G}{\partial A}\right)_{T,p,n_i} \qquad (4.8)$$

The chemical potential is defined as the increase in free energy of a system on adding an infinitesimal amount of a component (per unit number of molecules of that component added) when T, p and the composition of all other components are held constant. Clearly, from this definition if a component i in phase A has a higher chemical potential than in phase B (that is, $\mu_i^A > \mu_i^B$) then the total free energy will be lowered if molecules are transferred from phase A to B, and this will occur in a spontaneous process until the chemical potentials equalise at equilibrium. It is easy to see from this why the chemical potential is so useful in mixtures and solutions in matter transfer (open) processes. This is especially clear when it is understood that μ_i is a simple function of concentration, that is:

$$\mu_i = \mu_i^o + kTln_e C_i. \tag{4.9}$$

for dilute mixtures, where μ_i^o is the standard chemical potential of component i, usually 1 M for solutes and 1 atm for gas mixtures. This equation is based on the entropy associated with a component in a mixture and is at the heart of why we generally plot measurable changes in any particular solution property against the log of the solute concentration rather than using a linear scale. Generally, only substantial changes in concentration or pressure produce significant changes in the properties of the mixture. (For example, consider the use of the pH scale.)

THE GIBBS ADSORPTION ISOTHERM

Let us consider the interface between two phases, say between a liquid and a vapour, where a solute (i) is dissolved in the liquid phase. The real concentration gradient of the solute near the interface may look like that illustrated below:

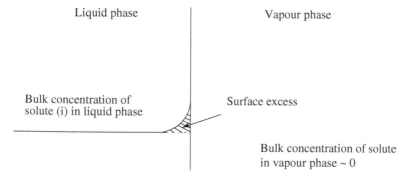

Figure 4.1 Diagram of the variation in solute concentration at an interface between two phases.

When the solute increases in concentration near the surface (e.g. a surfactant), there must be a surface excess of solute n_i^σ, compared with the bulk value continued right up to the interface. We can define a surface *excess* concentration (in units of moles per unit area) as:

$$\Gamma_i = \frac{n_i^\sigma}{A} \tag{4.10}$$

where A is the interfacial area (note that Γ_i may be either positive or negative).

Let us now examine the effect of adsorption on the interfacial energy (γ). If a solute i is positively adsorbed with a surface density of Γ_i, we would expect the surface energy to *decrease* on increasing the bulk concentration of this component (and the opposite to be true also). This situation is illustrated below, where the total free energy of the system G^T and μ_i are both increased by addition of component i, but because this component is favourably adsorbed at the surface (only relative to the solvent, since both have a higher energy state at the surface), the work required to create a new surface (i.e. γ) is reduced. Thus, although the total free energy of the system increases with the creation of a new surface, this process is made easier as the chemical potential of the selectively adsorbed component increases (i.e. with concentration).

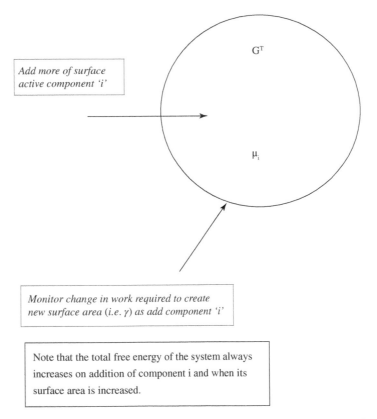

Figure 4.2   Diagram to illustrate the change in surface energy caused by the addition of solute.

This reduction in surface energy must be directly related to the change in chemical potential of the solute and to the amount adsorbed and is therefore given by the simple relationship:

$$d\gamma = -\Gamma_i d\mu_i \tag{4.11}$$

or, for the case of several components:

$$d\gamma = -\Sigma_i \Gamma_i d\mu_i. \tag{4.12}$$

The change in μ_i is caused by the change in bulk solute concentration. This is the Gibbs surface tension equation. Basically, these equations describe the fact that increasing the chemical potential of the adsorbing species reduces the energy required to produce new surface (i.e. γ). This, of course, is the principal action of surfactants, which will be discussed in more detail in a later section.

Using this result, let us now consider a solution of two components

$$d\gamma = -\Gamma_1 d\mu_1 - \Gamma_2 d\mu_2 \tag{4.13}$$

and hence the adsorption excess for one of the components is given by:

$$\Gamma_1 = -\left(\frac{\partial \gamma}{\partial \mu_1}\right)_{T,\mu_2} \tag{4.14}$$

Thus, in principle, we could determine the adsorption excess of one of the components from surface tension measurements *if* we could vary μ_1 independently of μ_2. But the latter appears not to be possible, because the chemical potentials are dependent on the concentration of each component. However, for dilute solutions the change in μ for the solvent is negligible compared with that of the solute. Hence, the change for the solvent can be ignored and we obtain the simple result that:

$$d\gamma = -\Gamma_1 d\mu_1 \tag{4.15}$$

Now, since $\mu_1 = \mu_1^\circ + RTlnc_1$, differentiation with respect to c_1 gives:

$$\left(\frac{\partial \mu_1}{\partial c_1}\right)_T = RT\left(\frac{\partial \ln c_1}{\partial c_1}\right)_T = \frac{RT}{c_1} \tag{4.16}$$

then substitution in equation (4.15) leads to the result:

$$\Gamma_1 = -\frac{1}{RT}\left(\frac{\partial \gamma}{\partial \ln c_1}\right)_T = \frac{c_1}{RT}\left(\frac{\partial \gamma}{\partial c_1}\right)_T \qquad (4.17)$$

This is the important Gibbs adsorption isotherm. (Note that for concentrated solutions, the activity should be used in this equation.) Experimental measurements of γ over a range of concentrations allow us to plot γ versus $\ln c_1$ and hence obtain Γ_1, the adsorption density at the surface. The validity of this fundamental equation of adsorption has been proven by comparison with direct adsorption measurements. The method is best applied to liquid/vapour and liquid/liquid interfaces, where surface energies can easily be measured. However, care must be taken to allow equilibrium adsorption of the solute (which may be slow) during measurement.

Finally, it should be noted that equation 4.17 was derived for the case of a single adsorbing solute (e.g. a non-ionic surfactant). However, for ionic surfactants such as CTAB, two species (CTA^+ and Br^-) adsorb at the interface. In this case the equation becomes:

$$\Gamma_1 = -\frac{1}{2RT}\left(\frac{\partial \gamma}{\partial \ln c_1}\right)_T \qquad (4.18)$$

because the bulk chemical potential of *both* ions change (equally) with concentration of the surfactant.

(Question: Consider which form of the isotherm would apply to an ionic surfactant solution made up in an excess of electrolyte.)

DETERMINATION OF SURFACTANT ADSORPTION DENSITIES

Typical results obtained for the variation in surface tension with surfactant (log) concentration for a monovalent surfactant are given below:

These results show several interesting features. At any point on the curve, the value $d\gamma/d\ln C$ gives, from the Gibbs adsorption equation 4.18, the corresponding value of the surfactant adsorption density or, alternatively, the surfactant head group or packing area at the water-air interface. As we will see in a later section, another method for determining the surfactant head group area is afforded by the Langmuir trough technique (see

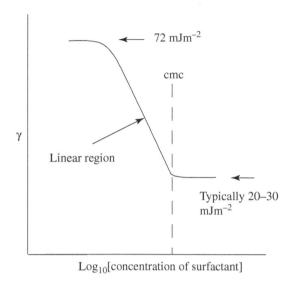

Figure 4.3 Typical experimental graph of measured surface energy versus concentration for a micelle-forming surfactant.

Chapter 10). At surfactant concentrations just below the cmc value, the slope $d\gamma/dlnC$ is linear, which from the Gibbs equation means that there is no further increase in the adsorption density, with increase in bulk concentration. The surface is fully packed with surfactant molecules, although γ still continues to fall. This apparently odd situation arises because the chemical potential of the surfactant continues to increase with its concentration (see equation 4.9) in this region, and although its adsorption density does not change, this must reduce the energy required to create new surface; hence the surface energy continues to fall.

However, at the cmc a sharp transition occurs which apparently corresponds to zero adsorption (i.e. $d\gamma/dlnC = 0$)! How can this be so? If we examine properties of the bulk solution in this region, we find that at this same concentration there is a sharp transition in a wide range of properties, such a conductivity, osmotic pressure and turbidity (see the following chapter). What is in fact happening is that the surfactant molecules are forming aggregates, usually micelles, and that all the additional molecules added to the solution go into these aggregates, so the concentration of monomers remains roughly constant. That is, both $d\gamma$ and $dlnC$ are effectively zero, and the plot should strictly stop at the cmc value, since although we are adding surfactant molecules, we are not increasing their activity or concentration. The precise nature of these aggregates is discussed in the following chapter.

Industrial Report

Soil microstructure, permeability and interparticle forces.

A soil is a condensed colloid system because the negatively charged plate-shaped crystals are assembled in parallel or near parallel alignment to form stable operational entities, described as clay domains. The crystals within a clay domain can be represented by a three-plate crystal model in which one crystal separates the other two crystals to produce a slit-shaped pore, where the crystals overlap. This situation is illustrated below:

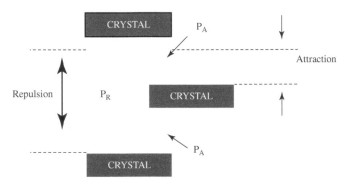

Figure 4.4 Schematic diagram of a pore that can be formed in clay crystal domains.

The surface separation in the slit-shaped pore is determined by the crystal thickness. For an illite (a fine-grained mica with a surface area of 1.6×10^5 m^2 per kg), the slit-shaped pores have a median size of about 5 nm, and in the overlap pores the surface separation is about 1 nm. The stability of clay domains within a soil is a crucial feature for agricultural production, because the permeability of a soil to aqueous electrolyte solutions depends on this stability. Swelling of these domains reduces permeability.

The interaction of clay crystals within a domain depends upon the DLVO repulsive pressure in the slit-shaped pores and the balance between repulsive pressure (P_R) from counterion hydration and the attractive pressure (P_A) generated by van der Waals forces and the recently discovered ion-ion correlation attraction between the counterions in the confined space of the overlap pores (see: R. Kjellander et al. *J. Phys. Chem.*, 92, 6489–6492, 1988 and *J. Colloid & Interface Sci.*, 126, 194–211, 1988). When Ca^{2+} is the counterion, the attractive pressure dominates, and the overlap pores are stabilised in a primary potential minimum. However, when the crystal charge is balanced, in part, by Na^+ ions (called soil sodicity), the DLVO repulsive pressure in the slit-shaped pores increases, and the ion-ion correlation attraction is reduced. In dilute solutions the repulsive pressure in the slit-shaped pores is sufficient to release the platelets from the shallow potential minimum, and the domains start to swell. However, in accord with DLVO theory, the swelling pressure in the slit-shaped pores can be reduced by increasing the electrolyte concentration.

Studies on soils have shown that there is a nexus between saturated permeability (zero suction), sodicity and electrolyte concentration. The concentration, obtained by diluting the electrolyte, at which there is a first discernible decrease in permeability, called the threshold concentration, corresponds to the start of the swelling of the clay domains.

In irrigation agriculture this threshold concentration is used as a reference to adjust the concentration of irrigation water so that it exceeds the threshold concentration for the sodicity of the soil and so prevents decreases in permeability. The threshold concentration increases with the degree of sodicity (see J. P. Quirk, *Aust. J. Soil Res.*, 39, 1185–1217, 2001).

At about one-quarter of the threshold concentration for a given sodicity, dispersed particles appear in the percolate, indicating the start of the dismantling of clay domains. It is noteworthy that this concentration is almost ten times lower, or even more if natural dispersants are present (e.g. organic compounds), than that obtained for the flocculation of a suspension of the soil. This reflects the fact that it is harder to release the crystals from within the clay domains than to simply flocculate the free crystals.

Professor J. P. Quirk
Formerly Director
Waite Agricultural Research Institute
Adelaide, Australia

SAMPLE PROBLEMS

1. The surface tension data given in the following figure was obtained for
    aqueous solutions of a trivalent cationic surfactant ($CoRCl_3$) in both
    water and in 150 mM NaCl solution. Use the data and the Gibbs adsorp-
    tion isotherm to obtain estimates of the minimum surface area per mol-
    ecule adsorbed at the air/water interface.

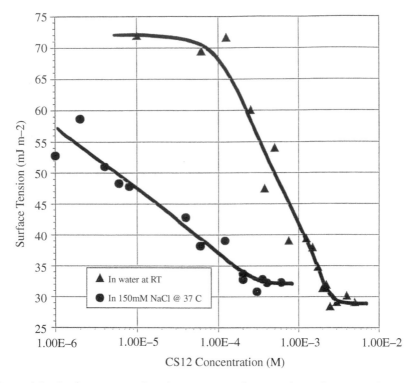

Figure 4.5 Surface tension data for aqueous solutions of a surfactant with a triva-
lent headgroup $CoR^{3+} Cl_3$.

2. Use the Gibbs adsorption isotherm to describe the type of surfactant
    adsorption occurring at the air/water interface at points A, B, C and D
    on the following graph.

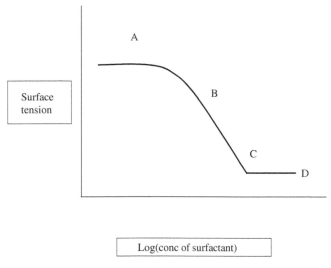

Figure 4.6 Schematic diagram of the decrease in surface tension with concentration for a typical surfactant solution.

Experiment

Adsorption of Acetic Acid onto Activated Charcoal

Introduction

Activated granular charcoal or carbon is widely used for vapour adsorption and in the removal of organic solutes from water. These materials are used in industrial processes to purify drinking and swimming pool water, decolorise sugar solutions as well as other foods and extract organic solvents. It also used as a first (oral) treatment in hospitals for cases of poisoning. Activated charcoal can be made by heat degradation and partial oxidation of almost any carbonaceous material of either animal, vegetable or mineral origin. For convenience and economic factors, it is usually produced from bones, wood, lignite and coconut shells. The complex three-dimensional structure of these materials is determined by their carbon-based polymers (such as cellulose and lignin), and it is this backbone which gives the final carbon structure after thermal degradation. These materials, therefore, produce a very porous high surface area carbon solid. In addition to a high area, the carbon has to be 'activated' so that it will interact with and physisorb a wide range of compounds. This activation process involves controlled oxidation of the surface to produce polar sites. In this experiment, we will examine quantitatively the adsorption properties of a typical granular charcoal. Adsorption at liquid surfaces can be monitored using the Gibbs adsorption isotherm since the surface energy of a solution can be readily measured. However, for solid substrates this is not the case, and the adsorption density has to be measured in some other manner. In the present case, the concentration of adsorbate in solution will be monitored. In place of the Gibbs equation, we can use a simple adsorption model based on the mass action approach.

If we assume that the granular charcoal has a certain number of possible adsorption sites per gm (N_m) and that a fraction θ are filled by the adsorbing solute, then:

the rate of adsorption α [solute conc.]$[1-\theta]N_m$

and

the rate of desorption α θN_m

At equilibrium these rates must be equal, hence

$$k_a C(1-\theta)N_m = k_d \theta N_m \qquad (1)$$

where k_a, k_d are the respective proportionality constants and C is the bulk solution concentration of solute. If we let $K = k_a/k_d$, then

$$C/\theta = C + 1/K \qquad (2)$$

and since $\theta = N/N_m$, where N is the number of solute molecules adsorbed per gm of solid, we obtain the result that:

$$\frac{C}{N} = \frac{C}{N_m} + \frac{I}{KN_m} \qquad (3)$$

Thus, measurement of N for a range of concentrations (C) should give a linear plot of C/N versus C, where the slope gives the value of N_m and the intercept the value of the equilibrium constant K.

This model of adsorption was suggested by Langmuir and is referred to as the Langmuir adsorption isotherm. The aim of this experiment is to test the validity of this isotherm equation and to measure the surface area per gm of charcoal, which can easily be obtained from the measured N_m value if the area per solute molecule is known.

Experimental procedures

In this experiment it is important to measure the acetic acid concentrations accurately. To this end, the NaOH solution must be titrated with a standard 0.1 M HCl solution and then titrated with the acetic acid solution. (Question: why is NaOH solution not used as a standard?)

Weigh out 1 gm of granulated activated charcoal into each of 5 clean stoppered 250 cm^3 conical flasks. Add 100 cm^3 of 0.2 M acetic acid (stock solution) to the first flask and shake. Then add 100 cm^3 to each of the other flasks at concentrations of: 0.15, 0.10, 0.07 and 0.03 M. Shake each loosely stoppered flask periodically over 30 minutes. Then allow the flasks to stand for one hour, and note the room temperature. Withdraw just over 50 cm^3 of the solution and filter through a fine sinter (to completely remove charcoal particles) and titrate two 25 cm^3 portions with 0.1 and 0.01 M NaOH (depending on the initial acid concentration). As an indicator use phenolphthalein. It is important that the equilibrium acetic acid concentration be accurately determined.

Calculate the number of acetic acid molecules adsorbed per gm of charcoal (N) and the corresponding *equilibrium* acid concentration (C). Plot N versus C and C/N versus C. Determine the surface area per gm of charcoal, assuming that one adsorbed acetic acid molecule occupies an area of 21 $Å^2$. Estimate the value of the equilibrium constant K with the correct units.

FOR CONSIDERATION/ TYPICAL QUESTIONS

1 The area per gm of activated granular charcoal as determined by nitrogen adsorption is typically between 300–1000 m^2/gm. Why is this value different to that determined in this experiment?

2 Why should the carbon surface be activated by oxidation, and what would you expect the typical surface groups to be?

3 Why is a slurry of charcoal in water given orally to suspected poison victims?

5

Surfactants and Self-Assembly

Introduction to the variety of types of surfactants, effect of surfactants on aqueous solution properties. Law of mass action applied to the self-assembly of surfactant molecules in water. Spontaneous self-assembly of surfactants in aqueous media. Formation of micelles, vesicles and lamellar structures. Critical packing parameter. Detergency. Laboratory project on determining the charge of a micelle.

INTRODUCTION TO SURFACTANTS

The name 'surfactant' refers to molecules which are 'surface active', usually in aqueous solutions. Surface active molecules adsorb strongly at the water-air interface, and because of this, they substantially reduce its surface energy (Gibbs theorem). This is the opposite behaviour to that observed for most inorganic electrolytes, which are desorbed at the air interface and hence raise the surface energy of water (slightly). Surfactant molecules are amphiphilic; that is, they have both hydrophilic and hydrophobic moieties, and it is for this reason that they adsorb so effectively at interfaces (note that amphi means 'of both kinds' in Greek).

Natural surfactants, such as soaps, are made by saponification of fats or triglycerides, such as tripalmitin in palm oil. The main component of common soap is sodium stearate $C_{17}H_{35}COO^-$ $Na^\oplus$, which is made from the saponification of animal fats. When dissolved in water, the carboxylic head-group ionises and is strongly hydrophilic, whereas the hydrocarbon chain is hydrophobic. The hydrocarbon chain, alone, is almost completely insoluble

Applied Colloid and Surface Chemistry, Second Edition. Richard M. Pashley and Marilyn E. Karaman.
© 2021 John Wiley & Sons Ltd. Published 2021 by John Wiley & Sons Ltd.
Companion website: www.wiley.com/go/pashley/appliedcolloid2e

in water. When dissolved into aqueous solution, the molecules can adsorb and orientate at the air/solution interface, as illustrated below, to reduce the surface tension of water:

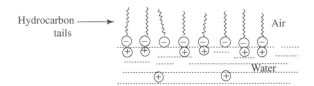

Figure 5.1   Schematic diagram of surfactant molecules adsorbed at the water/air interface.

The interfacial energy is typically reduced down to about 30–35 mJm^{-2}, and the surface now appears more like that of a hydrocarbon. The surfactant molecules are in a lower energy state when immersed in bulk water than when adsorbed at the surface. However, the displacement of even less favourable water molecules from the air surface dominates the overall process, and the surfactant molecules are preferentially adsorbed. It is the hydrocarbon tail which makes the molecule less favourable in water. The methylene groups can neither hydrogen bond nor form dipole bonds with water. Water molecules surrounding the hydrocarbon groups therefore orientate or order so as to increase the number of bonds to other neighbouring water molecules. This increase in local order (decreasing entropy) increases the free energy of these water molecules relative to those in bulk solution. A list of typical surfactant molecules, with different types of charge, is given below:

Table 5.1

ANIONIC:	
Sodium dodecyl sulphate (SDS)	$CH_3(CH_2)_{11}SO_4^-Na^+$
Sodium dodecyl benzene sulphonate	$CH_3(CH_2)_{11}C_6H_4SO_3^-Na^+$
CATIONIC:	
Cetyltrimethylammonium bromide (CTAB)	$CH_3(CH_2)_{15}N(CH_3)_3^+Br^-$
Dodecylamine hydrochloride	$CH_3(CH_2)_{11}NH_3^+Cl^-$
NON-IONIC:	
Polyethylene oxides	e.g. $CH_3(CH_2)_7(OCH_2CH_2)_8OH$
	(called C_8EO_8)

Table 5.1   (*Continued*)

ZWITTERIONIC:

Dodecyl betaine

$$C_{12}H_{25}N^+\Big\langle{}^{(CH_3)_2}_{CH_2COO^-}$$

Lecithins, e.g phosphatidyl choline

$$CH_2OCR \overset{O}{\overset{\|}{}}$$

$$CHOCR \overset{O}{\overset{\|}{}}$$

$$CH_2OP\text{-}O\text{-}CH_2\text{-}\overset{+}{N}(CH_3)_3 \overset{O}{\overset{\|}{}}$$

$$O_-$$

In addition to the surface adsorption properties of surfactants, they also have the remarkable ability to self-assemble in aqueous solution. The structures spontaneously formed by surfactants in solution are created to reduce the exposure of the hydrocarbon chains to water. Many of their solution properties reflect this ability, as is illustrated on the following diagram, which shows typical solution behaviour for single-chained ionic surfactants such as CTAB and SDS.

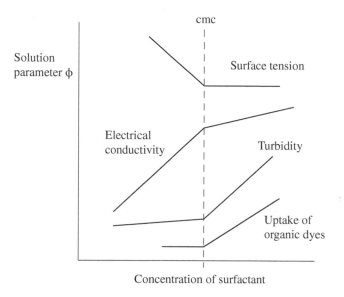

Figure 5.2   Diagram illustrating the sharp change in a range of solution properties at the cmc.

A sharp transition occurs in most of the solution properties, which corresponds to the formation of self-assembled structures called micelles. The concentration at which they are formed is a characteristic of the particular surfactant and is called the 'critical micelle concentration', or cmc. A section through a micelle would look something like the following schematic diagram:

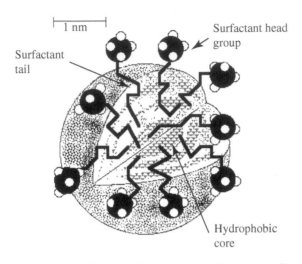

Figure 5.3   Schematic diagram of a surfactant micelle.

These small structures typically contain about 100 molecules, with the central core of the micelle essentially a water-free liquid hydrocarbon environment. The formation of aggregates is a very important property of surfactants and is of fundamental importance in their detergent cleaning action. The hydrocarbon regions in the aggregates solubilise fatty organic materials during cleaning. This organic 'dirt' is otherwise insoluble in water. As an example, if liquid paraffin is stirred into a soap solution, the solution remains clear until the capacity of the micelles to absorb paraffin is exceeded. Water-insoluble coloured organic dyes, such as Sudan yellow, are clearly taken up in micellar solutions, as shown in the following photograph. This illustrates the fundamental process we use every day for washing dirt from ourselves and our clothes etc.

THERMODYNAMICS OF SURFACTANT SELF-ASSEMBLY

That surfactant molecules form aggregates designed to remove unfavourable hydrocarbon-water contact is not surprising, but the question that should be asked is: Why do the aggregates form sharply at a concentration

Figure 5.4 Sudan yellow is a water-insoluble organic dye, seen at the bottom of the cylinder on the left, but is fully dissolved in the micelle solution on the right.

characteristic of the surfactant (i.e. at a specific concentration labelled the cmc)? From the basic equation of ideal solution thermodynamics:

$$\mu_i = \mu_i^\circ + kT ln x \tag{5.1}$$

it is clear that as we increase the concentration of the surfactant, x (the mole fraction) increases and so does the chemical potential of the molecules, μ_i. (Note that this equation was written earlier, in Chapter 4, in terms of molar concentrations, with an appropriate adjustment in standard state.) If we make the pseudo-phase approximation that the micelles can be considered as a species like the surfactant monomers, we also obtain a similar relationship:

$$N\mu_N = N\mu_N^\circ + kT ln\left(x_N / N\right) \tag{5.2}$$

where $N\mu_N$ is the chemical potential of an aggregate of N surfactant molecules and x_N is the mole fraction of surfactant molecules in the N-aggregates. Now, at equilibrium $\mu_1 = \mu_N$, the chemical potentials of free monomers and monomers in the aggregates must be the same. Thus we obtain the result that:

$$\mu_i^\circ + kT ln x_1 = \mu_N^\circ + \left[kT ln\left(x_N / N\right)\right] / N \tag{5.3}$$

Equation (5.3) can be rearranged to show that:

$$\frac{x_1^N}{x_N} = a\,\text{constant}\,(\text{for a fixed value of } N)$$

(5.4)

This is simply the result we would expect from the law of mass action. Thus, for the reaction N monomers $\Leftrightarrow$ 1 micelle, we can immediately write that there is an equilibrium (association) constant given by:

$$\frac{x_{mic}}{x_1^N} = K_e$$

(5.5)

or

$$\frac{x_N}{Nx_1^N} = K_e$$

(5.6)

Now, if we define the cmc as that mole fraction of surfactant at which $x_N \cong x_1 = x_{cmc}$, then at the cmc:

$$\frac{x_{cmc}^{(1-N)}}{N} = K_e$$

(5.7)

In addition, if this definition of the cmc is also incorporated into equation (5.3), then we obtain the result that for large values of N (e.g. 100):

$$ln(x_{cmc}) = \frac{\mu_N^\circ - \mu_1^\circ}{kT}$$

(5.8)

Equations (5.7) and (5.8) are both very useful when analysing surfactant aggregation behaviour and experimental cmc values. The difference in standard chemical potentials $\mu_N^\circ - \mu_1^\circ$ must contain within it the molecular forces and energetics of formation of micelles, which could be estimated from theory. These will be a function of the surfactant molecule and will determine the value of its cmc.

The above analysis actually refers to the simplest case of aggregation of non-ionic surfactants. For monovalent ionic surfactants, the aggregation reaction becomes:

$$N \text{ ionic monomers} + Q \text{ counterions} \Leftrightarrow \text{micelle}$$

because the ionic monomers will usually be fully ionised, but the high electric field strength at the surface of the micelles will often cause adsorption of some proportion of the free counterions. Thus, the micelles will have a total ionic charge of $(\pm)$ $(N-Q)$. The equilibrium constant is, in this case, given by:

$$K_e = \frac{x_{mic}}{x_1^N x_{ion}^Q} \tag{5.9}$$

Calculated concentrations, using equation (5.9), for the various components, surfactant monomers, counterions and micelles for the case of CTAB micellisation (with a cmc of 0.9 mM) are shown below. Clearly, the micelle concentration increases rapidly at the cmc, which explains the sharp transition in surfactant solution properties referred to earlier. It is also interesting to note that the law of mass action (in the form of equation 5.9) predicts an increase in counterion (Br$^-$ ions) concentration and a decrease in free monomer concentration above the cmc.

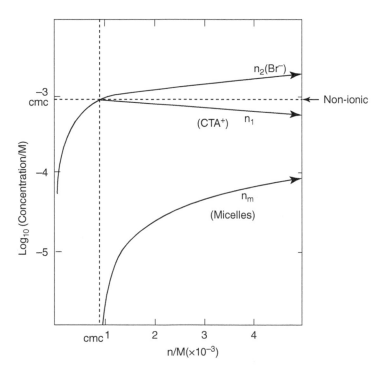

Figure 5.5 Calculated concentrations of micelles, CTA$^+$ and Br$^-$ ions for CTAB surfactant self-assembly near its cmc.

It has been proposed that for ionic surfactants, a useful definition of the cmc would be:

$$\frac{dn_1}{dc} = 0 \text{ at the cmc}$$

rather than the usual experimental definition of:

$$\frac{d^3\phi}{dc^3} = 0$$

which is difficult to use theoretically.

The equivalent curves for non-ionic surfactants, obtained using equations (5.6) and (5.7), give a constant monomer concentration above the cmc, with a similar increase in micelle concentration.

SELF-ASSEMBLED SURFACTANT STRUCTURES

Ionic surfactants actually only form micelles when their hydrocarbon chains are sufficiently fluid, that is at temperatures above their chain melting temperature. Below a specific temperature for a given surfactant, the Krafft temperature, the surfactant becomes insoluble rather than self-assembles. For CTAB this temperature is around 25 °C, and only above this temperature are micelles formed. In general, the longer the hydrocarbon chain length, the higher the Krafft temperature. For this reason, shorter chain length surfactants or branched chain soaps are used for cold water detergent formulations. Non-ionic surfactants suffer the inverse problem and become insoluble with increasing temperature. This is because their polar head-groups become less hydrophilic with increasing temperature and a 'cloud point' is reached where they precipitate from solution.

Micelles are not the only self-assembled structures that can be formed. The physical constraints on the surfactant molecule dictate the type of aggregate which it can form to exclude water. Possible structures can be modelled simply by using the surfactant's calculated hydrocarbon chain volume and its head-group area. These simple ideas can be used to illustrate clearly why double chain surfactants, such as the lecithins, do not form micelles but instead form vesicles, liposomes and multilamellar bilayers. These remarkable aggregates closely mirror the kinds of structures observed in living cell membranes. The chain volume varies with the surfactant type, but the addition of suitable chain-penetrating oils can increase this volume. An ionic surfactant head-group area can also be varied by the addition of screening electrolyte, which has the effect of reducing the head-group area and hence can dramatically change its aggregation properties. In addition, co-surfactants (usually long-chain alcohols) can be incorporated to change the packing. These can also lower the minimum surface energy at the air/ solution interface.

Molecular organisation or self-assembly depends upon a number of competing intramolecular forces, the flexibility of the chains and the intermolecular forces. The relative magnitudes of the attractive hydrophobic forces between the hydrophobic tails, the repulsive electrostatic forces between the charged head-groups and the head-group hydration effects all influence aggregate architecture and stability. Originally it was thought that in order to predict the physical characteristics of an aggregate (i.e. size, shape etc.), it would be necessary to have detailed knowledge of the complex intermolecular forces acting between the polar head-groups and their hydrocarbon tails. However, it soon became clear that simple packing constraints offered a valuable tool for the prediction of aggregate structure. This view of aggregation saw the emergence of a relatively simple characterisation of self-assembly based on the degree of curvature existing at the aggregate surface. This curvature can be expressed as a dimensionless parameter known as the critical packing parameter $v/a_o l_c$, where v is the volume of the hydrocarbon chain (assumed to be fluid and incompressible) given below, l_c is the critical chain length, assumed to be approximately equal to l_{max} the fully extended chain length, n is the number of carbon atoms in the hydrocarbon chain, m is the number of hydrocarbon chains and a_o is the head-group area (which can be estimated using the Langmuir trough technique, see later).

$$v = (27.4 + 26.9\,n)\,m \qquad \text{(units Å}^3\text{)}$$
$$l_{max} = 1.5 + 1.265\,n \qquad \text{(units Å)}$$

The critical packing parameter can be used as a guide to the aggregate architecture for a given surfactant (as shown in the following table). Typical values and their corresponding aggregate structures are given below:

$$v/a_o l_c < 1/3 ....\text{Spherical micelles}$$
$$1/3 < v/a_o l_c < 1/2 ....\text{Poly dispersed cylindrical micelles}$$
$$1/2 < v/a_o l_c < 1 ....\text{Vesicles oblate micelles or bilayers}$$
$$v/a_o l_c > 1 ....\text{Inverted structures}$$

Lipid	Critical packing parameter $\dfrac{v}{a_o l_c}$	Critical packing shape	Structures formed
Single-chained lipids with large headgroup areas. (eg. NaDS in low salt and some lysophospholipids.)	$< \dfrac{1}{3}$	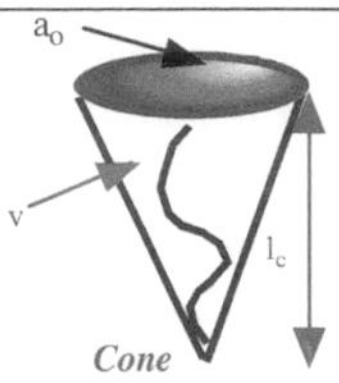 *Cone*	Spherical micelles 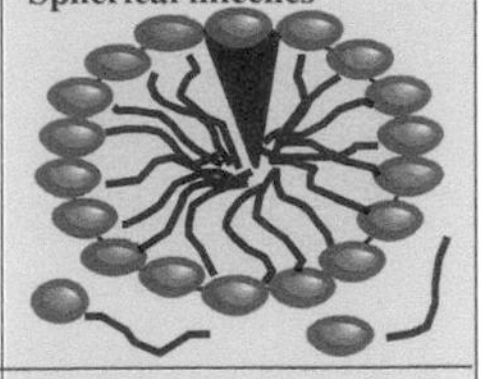
Single-chained lipids with small headgroup areas. (eg NaDS in high salt, lysolecithin and non-ionic surfactants.)	$\dfrac{1}{3} - \dfrac{1}{2}$	 *Truncated Cone or Wedge*	Cylindrical or Globular micelles 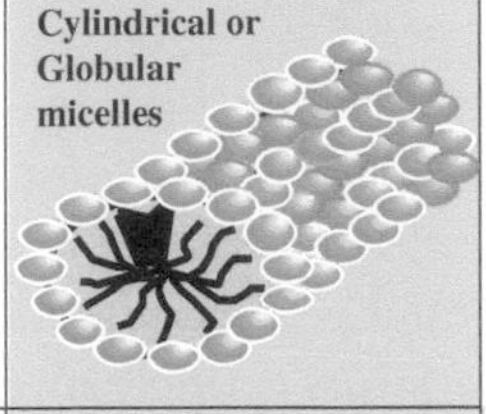
Double chained lipids with large headgroup and fluid chains. (eg Lecithin, dialkyl dimethyl ammonium salts, sphingomyelin, DGDG, phosphatidylserine, phosphatidyl inositol.)	$\dfrac{1}{2} - 1$	 *Truncated Cone*	Flexible Bilayers, Vesicles 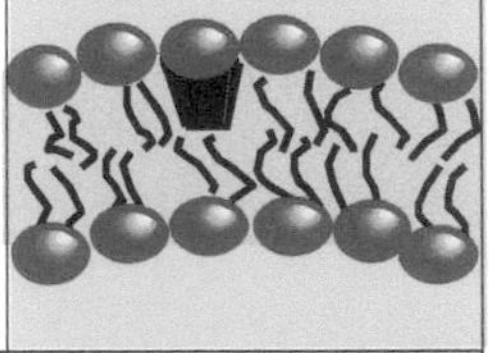
Double chained lipids with small head group areas: anionic lipids high salt, saturated frozen chains. (e.g phosphatidyl ethanolamine, phosphatidyl serine + Ca^{2+})	~ 1	 *Cylinder*	Planar Bilayers 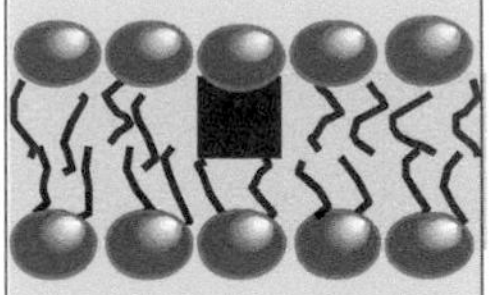
Double chained lipids having small head groups (eg. nonionic lipids, poly(cis) unsaturated chains, phosphatidic acid + Ca^{2+}.)	> 1	 *Inverted Truncated cone*	Inverted micelles 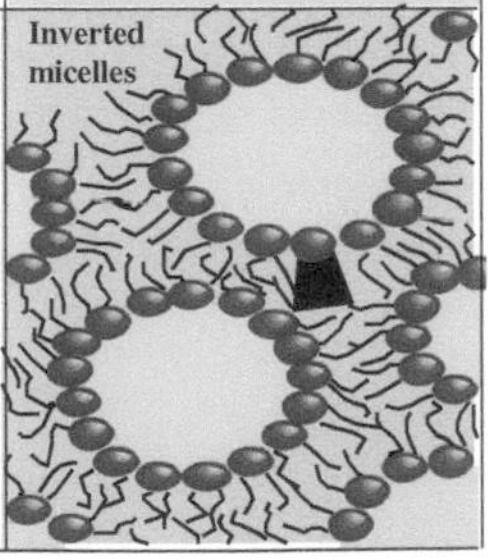

Figure 5.6 Use of the critical packing parameter to predict surfactant aggregate structures.

Earlier theories dealt with only these simple shapes; however, more complicated structures (e.g. cubic and other bicontinuous phases) predicted by geometric packing arguments have since been confirmed. The critical packing parameter is a useful parameter in aggregate design, as it can be changed for a given ionic surfactant by the addition of electrolyte, addition of co-surfactant, change in temperature, change in counter-ion or insertion of unsaturated or branched chains. Some of the basic structures are illustrated in the following table. See also the important text by Charles Tandon, *The Hydrophobic Effect* (New York: Wiley, 1980).

Controlling aggregate architecture has enormous potential in many new areas of biochemical research, catalysis, drug delivery and oil recovery, to name but a few.

Surfactants and detergency

In the detergency process, fatty materials (i.e. dirt, often from human skin) are removed from surfaces, such as cloth fibres, and dispersed in water. It is the surfactants in a detergent which produce this effect. Adsorption of the surfactant both on the fibre (or surface) and on the grease itself increases the contact angle of the latter, as illustrated in the following figure:

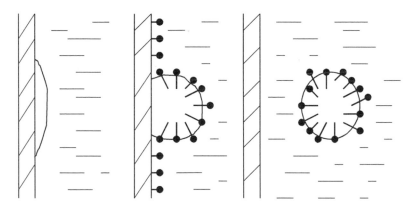

Figure 5.7 Illustration of the removal of hydrophobic oil from a fibre using detergent in water.

The grease or oil droplet is then easily detached by mechanical action, and the surfactant adsorbed around the surface of the droplet stabilises it in solution.

The composition of a typical powdered laundry detergent is given below:

Na alkyl sulfate soap	15%
Anhydrous soap	3%
Sodium tripolyphosphate/zeolite/polycarboxylates	30–50%
Sodium silicate	10–20%
Sodium carbonate	10%
Sodium sulphate	10–20%
Sodium carboxymethyl cellulose	1%
Optical brighteners, perfume, moisture, enzymes	~

The first two components are the active surfactants, whereas the other components are added for a variety of reasons. The polyphosphate chelates Ca^{2+} ions which are present (with Mg^{2+} ions also) in so-called hard waters and prevent them from coagulating the anionic surfactants. Concerns about phosphate build up in waste-water has reduced the use of these materials. Sodium silicate is added as a corrosion inhibitor for washing machines and also increases the pH. The pH is maintained at about 10 by the sodium carbonate. At lower pH values, the acid forms of the surfactants are produced, and in most cases these are either insoluble or much less soluble than the sodium salt. Sodium sulphate is added to prevent caking and ensures free-flowing powder. The cellulose acts as a protective hydrophilic sheath around dispersed dirt particles and prevents redeposition on the fabric. Foam stabilisers (non-ionic surfactants) are sometimes added to give a visible signal that sufficient detergent has been added. However, the creation of a foam is not necessary for detergent action in a conventional washing machine. A lather is, however, of value for personal washing, as it provides a mobile concentrated soap solution on the skin. In the 1970s the widespread use of non-biodegradable synthetic detergents led to extensive foaming in rivers and lakes and the consequent death of aquatic life through lack of aeration of the water. Great care is now taken to use readily biodegradable soaps.

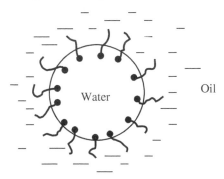

Figure 5.8   Diagram of how surfactant molecules can stabilize water droplets in oil.

An inverted type of detergency process is used in dry cleaning, where a non-aqueous liquid, usually tetrachloroethylene, is used to dissolve grease from materials that should not be exposed to water. However, because complete cleaning must also remove polar materials, for example sugars, which are insoluble in C_2Cl_4, surfactants and a small amount of water are added which form inverted micelles:

Polar materials are then solubilised inside the aqueous core of the micelle, and the material is not exposed to water, even though the process is not entirely dry. Mixtures of oils with different surfactants can be used to produce sticky inflammable gels such as napalm, made from palmitic soaps and aluminium salts. When water is replaced by a concentrated solution of a soluble oxidant, such as ammonium nitrate, the intimate contact with a reactant, such as hydrocarbon oil, can be used as the basis of a plastic explosive.

Finally, surfactants have also been used to reduce water evaporation from open reservoirs in arid areas, especially in Australia. The packed insoluble monolayer adsorbed at the air/water interface substantially reduces the transfer of water vapour to the atmosphere. Cetylalcohol is used at the rate of 1 oz per acre per day for this reason. It has been calculated that this procedure can save up to one million gallons per acre per year.

Industrial Report

Colloid science in detergency

The removal of particulate soil from a fabric surface is a key step in the overall detergency process, with the balance between van der Waals and electrical double layer forces playing a major role. However, once the particle has been removed from the fabric surface, keeping it from redepositing at some other stage in the process is vital if the article is to remain clean at the end of the wash process. Consequently, one needs to ensure that a barrier to deposition is established that will ideally persist throughout the process. So to ensure that redeposition does not occur, polymeric anti-redeposition agents (ARDs) are often added to a formulation. These ARD polymers work by adsorbing on both the substrate and the particulate soil, thereby creating a protective layer that can both sterically and electrostatically hinder redeposition of the previously removed soil. Typical ARD polymers are anionics such as sodium carboxymethyl cellulose (SCMC) and non-ionic cellulose ethers such as methylhydroxypropylcellulose (MHPC). Anionic polymers are particularly suitable for use with hydrophilic fabrics such as cotton and cotton mix, whereas non-ionics can be used on both polyester and cotton fabrics. Both types of polymer function by altering the surface properties of the fibres to render them more hydrophilic and by adsorbing at the particle and fabric surfaces to present a barrier via the absorbed layer to redeposition. SCMC is specific to hydrophilic fibres such as cotton and has best activity when the degree of substitution is below 0.7. In contrast, the non-ionic MHPC has a broader spectrum of activity, being effective on both hydrophilic and especially hydrophobic fibres such as polyester. Of course, in real detergent systems both surfactants and ARD polymers compete for the fabric and particle surfaces, and this competition can under certain circumstances reduce the polymer's effectiveness. This competition is seen to be most problematic with non-ionic surfactants as they compete for surface sites with the polymer. To ensure effectiveness under such conditions, the formulator needs to balance the surfactant and polymer levels carefully. Typical polymer levels are around 0.5% to 1.0% w/w in the formulation. Other polymer types that have been explored as ARDs are polyethylene/polyoxyethylene terephthalate copolymers and hydrophobically modified polyethylene glycol.

Dr Ian C. Callaghan
Head of Formulation Technology
Lever Faberge Europe – Global Technology Centre
Unilever Research
Vlaardingen
The Netherlands

SAMPLE PROBLEMS

(1) Describe how the critical packing parameter for surfactant self-assembly can be used to describe the structure of typical biological lipid membranes.
(2) Explain the link between the critical packing parameter and the interaction forces between surfactant molecules in water.
(3) Use the cmc values of a homologous series of single-chained sodium sulfate surfactants given below to estimate the standard free energy of

transfer of a methylene ($-CH_2-$) group from an aqueous to a hydrocarbon environment.

Number of carbons in tail	12	14	16	18
cmc/mM	8.6	2.2	0.58	0.23

Experiment

Determination of Micelle Ionization

Introduction

In this experiment the degree of ionization of CTAB micelles is determined by measuring the *change* in slope of solution electrical conductivity (κ) versus total concentration (C) as the solution goes through the critical micelle concentration (cmc). That this information is sufficient to estimate the degree of ionization of the micelles formed at the cmc can be shown by the following simple analysis.

First of all, let us make the assumption that the overall solution conductivity (κ) is entirely due to the free Br– ions present in solution. That is, we assume that the conductivity due to the much larger CTA+ ions and the even larger micelles is negligible because of the size of these species (and hence large viscous drag). Thus we assume that:

$$\kappa \alpha\, C_{Br^-} \tag{1}$$

Hence, at CTAB concentrations (C) below the cmc ($C <$ cmc)

$$\kappa = AC \tag{2}$$

since the CTAB monomer is always fully ionized and A is some constant. By comparison, at concentrations above the cmc ($C >$ cmc), it must also be true that:

$$C_{Br^-} = \text{cmc} + \alpha\left(C - \text{cmc}\right) \tag{3}$$

where α is the degree of ionization of the micelle, which is assumed to be independent of concentration. Thus, above the cmc:

$$\kappa = A\big[\,\text{cmc} + \alpha\,(C - \text{cmc})\big] \tag{4}$$

Now, from equations (2) and (4) we can calculate the gradients above and below the cmc:

$$\left.\frac{d\kappa}{dc}\right|_{c>\text{cmc}} = \alpha A \tag{5}$$

and

$$\left.\frac{d\kappa}{dc}\right|_{c<\text{cmc}} = A \tag{6}$$

and so, obviously, the ratio of these slopes gives the degree of ionization of the micelles.

Experimental details

All electrical conductivity measurements should be carried out on solutions equilibrated in a thermostat bath at 30 °C. Conductivity values (i.e. κ in units of ohm^{-1} cm^{-1}) should be determined at 1000 Hz and over a concentration range of 0.0003 M to 0.003 M, taking at least four measurements on either side of the cmc (~0.001 M). Plot a linear graph of the results and determine the cmc of CTAB and the approximate degree of ionization of the micelles. Note: In using surfactant solutions, always try to prevent foaming by not shaking solutions too vigorously.

FOR CONSIDERATION/TYPICAL QUESTIONS

(1) Discuss the main assumptions used in the method described here to estimate the degree of ionization of micelles.
(2) What is the total charge on a CTAB micelle if the head-group area is 45 Å^2 and the micelle is 40 Å in diameter?

6

PFAS Contamination

PFAS compounds or perfluorinated alkyl substances are a class of fluorocarbon surfactants which have been widely used as foam fire suppressants and for fabric coating to produce hydrophobic surfaces to prevent spills and wetting over the last 60 years. However, their widespread use has become a problem because of their potential bio- toxicity and accumulation in human bodies partly because of their inert nature. Contaminated soils and groundwater, particularly around military aircraft bases, have become a world-wide issue. The main focus now is on developing processes to remove these materials.

BACKGROUND TO PFAS CONTAMINATION

Per- and polyfluoroalkyl substances (PFAS) are groundwater contaminants that are present in hundreds of military fire training areas around the world. The US Environmental Protection Agency has issued drinking water health advisories for two common PFAS, perfluorooctanoic acid and perfluorooctanesulfonic acid, and thus removal of PFAS from contaminated sites has now become a critical issue. Remediation of PFAS presents a difficult challenge. Source drinking water must be treated to remove PFAS, usually by adsorption onto granular activated carbon (GAC), which leads to the problem of how to dispose of or treat PFAS once it is captured. The most successful and widely used approach is incineration of PFAS. However, the exact chemistry that occurs during the incineration process remains poorly understood.

Applied Colloid and Surface Chemistry, Second Edition. Richard M. Pashley and Marilyn E. Karaman.
© 2021 John Wiley & Sons Ltd. Published 2021 by John Wiley & Sons Ltd.
Companion website: www.wiley.com/go/pashley/appliedcolloid2e

It is interesting that the introduction of new surfactant materials has a long history of causing serious environmental issues. In the early 1960s European countries became concerned about novel synthetic, non-biodegradable surfactants causing foaming in rivers, leading to fish death through lack of dissolved oxygen. This led to legislation related to monitoring of chemical persistency for compounds released to the environment. With this history, it is hard to understand why the widespread use of fluoroalkyl soaps was allowed.

Fluoroalkyl compounds are strongly hydrophobic and have been widely used as a fabric and furniture coating to prevent the absorption of water-based spills. The presence of the C-F bonds ensured biological inertness, but another related side effect has been long-term bio-accumulation. The precise effects of such long term bioaccumulation is still being debated and analysed. Clearly, however, the need to remove our dependence on these materials has been recognized. It has been estimated that the half-life of these compounds in the human body is of the order of about seven years and so this change will take time.

The most common PFAS compounds, perfluorooctanoic and perfluorooctanesulfonic acid, were widely used to create fire-fighting foams particularly to suppress flammable liquid fires, such as those encountered with airplane crashes. These fluorinated compounds are formed via electrochemical fluorination, from alkyl compounds reacted with HF and telomerization, where sections of the chains are combined to produce mixed alkyl and perfluoro compounds. These compounds have been used as the basis for 'aqueous film-forming foams' (or AFFF) by the US Navy in the 1960s and have since been widely used by military fire-fighters. The foams they produce are effective fire suppressants, they have low viscosity and spread over most hydrocarbon fuels, preventing exposure to oxygen and they also cool the fire. The task now facing governments across the world is to develop efficient methods for their removal from groundwater and soils.

SO HOW DO WE REMOVE PFAS COMPOUNDS FROM THE ENVIRONMENT?

The current issue faced around the world is to develop efficient, low-cost methods to remove residual PFAS compounds from contaminated soil and contaminated groundwater now that the further use of these compounds has been banned. Acceptable drinking water levels of PFAS compounds are set at low levels, in the United States at <0.07 ppb, whereas contaminated groundwater can have levels of 10–20 ppm! In addition, there are thousands of PFAS molecules because of variation in chain length, oxidation, side chains and level of fluorination.

The recalcitrant nature of PFAS primarily results from the strength of the covalent bond (ca. 450 kJ/mole) between carbon and fluorine, as well as F-atom shielding effects. As a result, advanced oxidation processes (AOP) typically used in water and wastewater treatment are of limited value because hydroxyl radicals (OH•) exhibit minimal reactivity with PFAS.

Peroxides and TiO_2/UV methods have recently been tested. As an example, 99% of PFOS was destroyed following incineration at 600 °C, as confirmed by liquid chromatography–mass spectrometry (LC/MS) measurements. The consensus appears to be that the terminal functional group (e.g. the $-SO_3H$, $-COOH$ and related acid groups) can be cleaved off, resulting in a range of potentially toxic perfluoroalkenes and HF.

Adsorption to granular activated carbon and ion exchange are treatment approaches commonly applied by the US Department of Defense to clean water contaminated with per- and polyfluoroalkyl substances (PFASs). Thermal high-temperature decomposition is then typically considered as the next stage. But there are serious issues caused by unknown reactions and potentially toxic by-products to address. If using thermal degradation, it is probably best to use zeolites or modified zeolites. These act as adsorbents and catalysts for thermal decomposition and mineralization and can withstand high temperatures.

A SURFACE CHEMISTRY APPROACH

Since most PFAS compounds have either a sulfonic acid or carboxylic acid group on a linear or branched chain, they are essentially a mixture of strong and weak acid anionic surfactants. At normal drinking/groundwater pH, around 8, they will be in the typically Na^+ salt form. A simple approach being considered is therefore to use similar chain length surfactants as a type of co-surfactant to assist in the capture of these molecules at the surface of bubbles in a foam fractionation process. Once lifted off, the foam can be destabilized using silicone oil, ethanol or vacuum suction and then, if required, the carrier surfactant can be extracted and reused.

It seems that this approach has been slow to take off, in part perhaps because of the categorization of perfluorinated compounds as oleophobic – that is, not capable of mixing with hydrocarbons. It is true that fluorocarbon surfaces, such as Teflon frying pans, are oleophobic, and hydrocarbon oils do not wet the surface but form beaded droplets not unlike water on Teflon. This wetting behavior occurs because the van der Waals forces between hydrocarbon and fluorocarbon molecules are significantly weaker than between hydrocarbon and hydrocarbon. Hence, hydrocarbon droplets prefer to bead up rather than spread out on Teflon surfaces, giving them an oleophobic character. A comparison of the wetting properties of water and tetradecane on Teflon is shown in the following photo. The water droplet is on the left and illustrates the fact that water cannot form hydrogen bonds with Teflon. This is also consistent with the very low surface tensions of fluorocarbon liquids, of about 12 mJm^{-2}, which relates directly to their weak intermolecular van der Waals forces, and also when compared with alkanes, at about 18 mJm^{-2}.

Figure 6.1   Photograph of a water droplet (left) and a droplet of tetradecane on a Teflon solid.

This observation also demonstrates the dual oleophobic and hydrophobic nature of fluorocarbon materials. However, it should be carefully noted that neither fluorocarbon nor hydrocarbon chains can form hydrogen bonds with water molecules, so both are hydrophobic. They will therefore be forced together, when immersed in water, through the attractive hydrophobic interaction, which is driven by the ice-like water molecules, which are formed around the chains, being released into the bulk water. The hydrophobic attraction is actually caused by the reduction in free energy produced when water molecules are released back into their bulk phase.

That is, when immersed in water, fluorocarbon and hydrocarbon chains are forced together into molecular contact, so a suitable hydrocarbon carrier surfactant will attract a fluorocarbon surfactant and be readily adsorbed at the surface of a rising bubble and thereby be separated from solution, as illustrated schematically in the following molecular diagram.

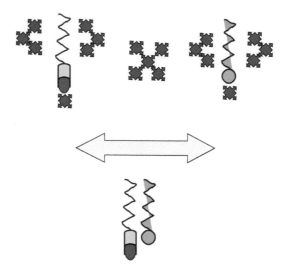

Figure 6.2 Schematic diagram of the origin of the hydrophobic attraction between fluorocarbon and hydrocarbon chains in water.

The use of a suitable carrier surfactant can therefore assist in the removal of PFAS compounds from groundwater in a simple physicochemical flotation process. Removal from soils is obviously more difficult, but a surfactant process might still work. This is because negatively charged PFAS compounds are most likely attached to negatively charged soil particles, generally composed of quartz, clay minerals and humic acid materials, by binding with Ca^{2+} ions, which is the most common metal ion found in soils. Hence, if we select an alkyl surfactant of similar chain length and head group chemistry but with the additional ability to bind Ca^{2+} ions, we could use a foam fractionation process to lift soil and release the adsorbed PFAS compounds, which can be carried to the surface.

One such potential surfactant is described in the following schematic diagram. In this case the cysteine headgroup is used to chelate the Ca^{2+} ion and so release the PFAS compound to be carried away in the foam. The cysteine surfactant acts also as the carrier compound. The collected foam can be broken up using a vacuum system or a silicone spray. The small quantities of PFAS concentrate could then simply be encased in concrete and buried.

This surfactant has the further advantage that it is readily decomposed into octanoic acid and cysteine, both of which are common food additives and so are environmentally benign. Alternatively, if required the surfactant could be extracted and reused.

The fact that PFAS compounds are typically present as low-level contaminants means that it would be more economic to use a suitable surfactant as a combined ion exchange/hydrophobic binding compound in a simple extraction/foam flotation process. The surfactant could combine selective ion exchange (for the release of Ca^{2+} bound PFAS compounds) and hydrophobic binding properties to support a low-energy foam fractionation process for soil remediation. This process is analogous to that currently used by industry, that is, granular activated charcoal (GAC) combined with Calgon for the ion exchange component. (Note that Calgon used to be based on the chelating polyphosphate ion, but now a mixture of zeolites and polycarboxylates is used.) A suitably designed surfactant can have the same action and in addition supports foam fractionation to separate the released PFAS compounds. A simple schematic diagram of how the surfactant could work is shown below.

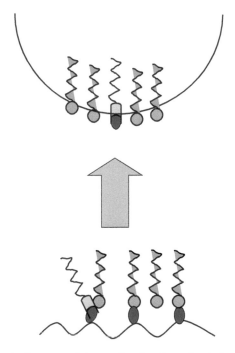

Figure 6.3 Schematic diagram of the mode of action of a chelating surfactant (with green head group) detaching a PFAS compound (in blue shading) from a soil particle held by a bridging Ca^{2+} ion (red) in the first stage of a foam fractionation soil remediation process.

The foam produced can then be easily collapsed/condensed, and the surfactant extracted and reused (using acid treatment and/or suitable hydrocarbon solvent). The almost pure PFAS component can then simply be encapsulated in concrete and stored safely for many years. This offers a more cost-efficient method compared with granular activated charcoal or IEX resins combined with high-temperature thermal decomposition.

Why bother to use high-temperature mineralization when we can safely encapsulate pure PFAS in a suitable concrete for long-term storage? Using a suitable surfactant as the basis for a simple physicochemical process for PFAS removal could then produce fairly pure PFAS for safe storage and disposal.

Experiment 6.1   Co-flotation removal of PFAS compounds from contaminated water.

Introduction

Almost all PFAS compounds have surface active properties, and it is therefore reasonable to expect that surfactants and typical surfactant separation processes could play a role in their removal from contaminated water (e.g. groundwater around airports) and even from soils.

In this experiment, for reasons of safety, one of the common soaps sodium octanate or sodium octylsulfonate (see below) is used as a substitute for the two most common PFAS contaminants, perfluorooctanoic acid and perflorooctanesulfonic acid.

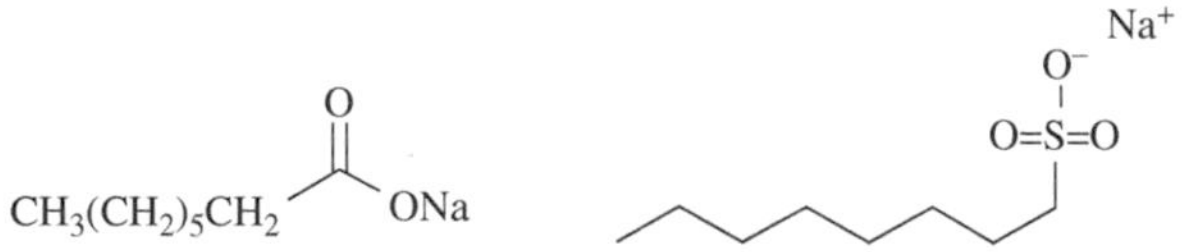

Figure 6.4   Sodium octanate or sodium octylsulfonate.

The dilute solutions of these two acids in drinking water or groundwater at pH 8 means that the acids (as well as the PFAS acids) will be present in the Na+ salt form.

In this experiment, for simplicity and the rapid production of results, the electrical conductivity of all solutions will be measured under N_2 cylinder sparging to remove dissolved atmospheric CO_2 gas. This will facilitate accurate values for the concentrations of the model contaminants shown in Figure 6.1, which will be monitored during the separation process.

The co-flotation process used for PFAS compounds removal in this experiment depends on the use of a flotation column with either non-ionic or zwitterionic surfactants added, in excess, which will not significantly change the electrical conductivity of the solution, so that the removal of the PFAS model compounds can be monitored using electrical conductivities.

Suitable compounds to be used for co-flotation are given below:

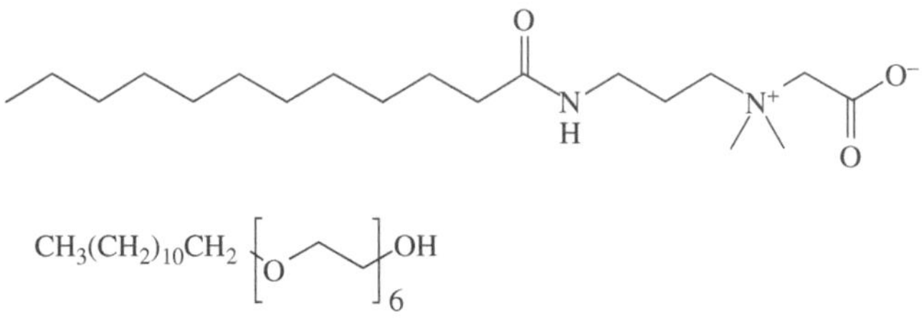

Figure 6.5 Cocamidopropyl betaine or lauramidopropyl betaine and hexaethylene glycol monododecyl ether.

Either of these surfactants can be used for the co-flotation experiments, since they will not significantly affect the electrical conductivity of the flotation solutions.

Experimental details

The model PFAS solutions will initially have a concentration level of 40 ppm to imitate severely contaminated groundwater, and the pH of the solutions should be increased by adding droplets of 0.01 M NaOH to increase the pH of the nitrogen sparged solutions to about 8.

This solution is then added to a nitrogen flotation tube as illustrated in the following figure.

The column should initially be filled with a solution of the surfactant carrier at a concentration of about 100 ppm and at pH 8. The top of the column should be connected via a Quickfit glass flange joint to allow addition of the model PFAS compound in the form of a small volume of concentrated solution to be released via pipette to just above the sinter, to allow rapid mixing. The top can also be removed

Figure 6.6   Suitable foam/bubbling flotation tube with pore size 2 glass sinter.

to enable solution samples to be removed from the column regularly for conductivity measurements. Note that it is important that a slow rate of N_2 gas sparging should be used in the conductivity measurement vessel to exclude atmospheric CO_2.

Ideally the N_2 gas flowrate should be controlled to produce a continuous foam overflow to efficiently remove the model PFAS compounds.

(If safety considerations can be satisfied, the experiment can be repeated using the perfluorinated compounds in place of the safer model hydrocarbons and the efficiency of separation compared.)

Your results should be plotted in the form of solution electrical conductivity vs time of flotation and also as the estimated concentration of the charged model compound as a function of flotation time.

If you have time, measure and construct a plot of electrical conductivity versus ppm concentration under a N_2 atmosphere for the selected surfactant carrier.

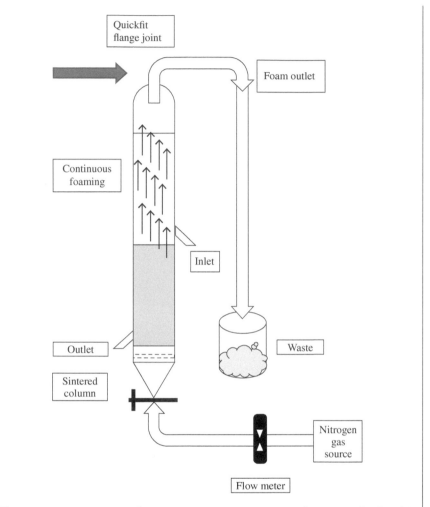

Figure 6.7 Foam or co-flotation separation apparatus for removal of PFAS model compounds.

QUESTIONS

(1) The advancing water contact angle on polypropylene is about 102° and on Teflon (PTFE) sheets about 110°, so why is the water contact angle just less than 90° on polyvinylidene fluoride (PVDF)?

(2) The advancing contact angle of the hydrocarbon liquid decane on Teflon is about 42°, which means that Teflon is oleophobic. How is this related to the strength of van der Waals forces between hydrocarbon and fluorocarbon molecules and with each other?

(3) Can you use their relative intermolecular van der Waals forces to explain why hydrocarbon and fluorocarbon chained surfactant molecules float together in water?

7

Emulsions and Microemulsions

The conditions required to form an emulsion of oil and water and a microemulsion. The complex range of structures formed by a microemulsion fluid. Emulsion polymerization and the production of latex paints. Photographic emulsions. Emulsions in food science and explosives. Laboratory project on determining the phase behaviour of a microemulsion fluid.

It is a well known that oil and water don't mix, and this can easily be demonstrated using a coloured light hydrocarbon oil and water, which will immediately phase separate even after vigorous shaking.

THE CONDITIONS REQUIRED TO FORM EMULSIONS AND MICROEMULSIONS

Adsorption of a surfactant monolayer at the air/solution interface still produces a surface with a significant interfacial energy, of typically between 30 and 40 mJm^{-2}, because the surface is now similar to that of a typical hydrocarbon liquid. However, in comparison, adsorption at a water/oil interface offers the opportunity of creating surfaces with very low interfacial energies, which ought to produce some interesting effects and possibilities. The formation of this type of low interfacial energy surface is the basis of the stability of most oil and water emulsions and all microemulsions. This situation is illustrated below:

Applied Colloid and Surface Chemistry, Second Edition. Richard M. Pashley and Marilyn E. Karaman.
© 2021 John Wiley & Sons Ltd. Published 2021 by John Wiley & Sons Ltd.
Companion website: www.wiley.com/go/pashley/appliedcolloid2e

Figure 7.1 Light hydrocarbon oil droplets, coloured by (blue) azulene dye, phase separate rapidly within a few seconds, even after vigorous shaking.

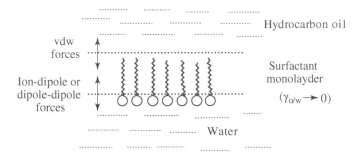

Figure 7.2 Illustration of the effect of an adsorbed surfactant layer on the interfacial energy between oil and water.

It is a well-known observation that oil and water do not mix. Fundamentally, this is because hydrocarbon molecules are non-polar and cannot interact strongly with water molecules, which have to be forced apart to incorporate the hydrocarbon solute molecules. Hence, hydrocarbon oils will not dissolve in water. However, when oil and water are vigorously shaken together, a droplet emulsion can be formed. This mixture will destabilize fairly quickly and phase separate into oil and water, typically within less than an hour, because of the high interfacial energy of the oil/water droplets.

However, the stability of these emulsions can be substantially enhanced by the addition of surfactants which will reduce the interfacial energy of the droplets. Emulsion droplets of oil in water or water in oil, typically fall in the range of 0.1–10 µm, whereas microemulsion droplets are in the 0.01–0.1 µm size range. However, in the latter case the actual structure of the phase is still very controversial (see later). For emulsions, the $\gamma_{o/w}$ value is typically in the range 0.1–1 mJm^{-2}, and for microemulsions $\gamma_{o/w}$ is as low as 0.001 mJm^{-2}. The addition of emulsifying agents (usually surfactant

Figure 7.3   Oil-in-water emulsion stabilized by the addition of surfactants.

mixtures) produces either an opaque stable emulsion or a clear microe-mulsion. This method of solubilising oil in water is used in foodstuffs (dairy produce), agricultural sprays and pharmaceutical preparations. A photograph of a relatively stable oil dispersion or emulsion in water is shown above. The droplet sizes typically of around 1 μm and the higher refractive index of oil produce the uniform light scattering effect or cloudy look.

Lowering the interfacial energy enables the formation of high-surface-area emulsions, but additional factors are also involved in preventing drop-let collisions from causing a phase separation. (Compare this situation with the coagulation of sols in Chapter 9.) These factors include electrical repul-sion for charged emulsifiers, polymers such as proteins adsorbed to give mechanical prevention of film drainage (and hence droplet coalescence), finely divided solid particles (such as clays and carbons) adsorbed around the interface and an increased viscosity. The type of emulsifier used deter-mines whether an oil-in-water or a water-in-oil emulsion is formed. An empirical numbering system has been developed to enable the correct type of surfactant to be chosen. The system is called the hydrophile-lipophile bal-ance, HLB. The most hydrophilic surfactants have the highest HLB values. In general, the phase in which the emulsifying agent is more soluble tends to be the dispersion medium.

Emulsions are metastable systems for which phase separation of the pure oil and water phases represents the most stable thermodynamic state. However, microemulsions, in which the interfacial energies approach zero, may be thermodynamically stable. Also, the microemul-sion phase is clear, which indicates very small droplets and a very high interfacial area. In fact, it is not at all certain what is the precise struc-tural nature of many microemulsions. It has been postulated that a range of intricate bicontinuous structures may exist, rather than simple drop-lets like swollen micelles. Whatever the nature of microemulsions, there is currently a large international effort at understanding these phases. One of the most important reasons for this is the possibility of using them to increase the yield from vast oil reserves in capillary rocks, which at present cannot be tapped. As an example, something like 95% of the anticipated oil wells have already been discovered in the United States., but the overall average recovery is less than 40%, so clearly a large amount of oil has yet to be removed.

The possible structures which can be formed by a mixture of hydro-carbon oil, surfactant and water are illustrated below:

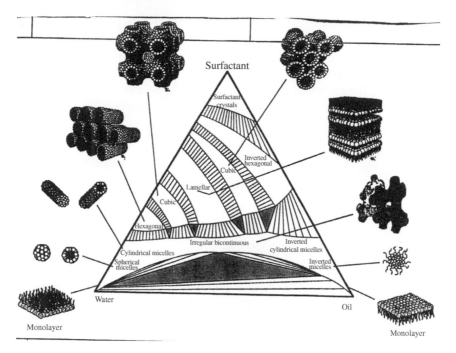

Figure 7.4 Schematic diagram of the types of structures formed at different compositions of oil, water and surfactants.

The variation and complexity of these structures have led to much research on potential industrial applications, from tertiary oil recovery to enhanced drug delivery systems. Many of the structures can be predicted using models based on the optimal curvature of the interface, not unlike that used to predict surfactant aggregation.

Emulsion or latex paints

The aim of the modern emulsion painting process is to deposit a uniform tough polymer layer on a substrate. At first sight it may be thought that simply dissolving the polymer (typically polyacrylic) in a suitable non-aqueous solvent would be sufficient. However, polymer solutions of the required concentrations (~50%) are very viscous, and organic solvents are not acceptable for the home decoration market as well as being expensive.

Einstein derived a simple equation for the viscosity of a solution of spherical particles, and from this result it is obvious that if we could make the polymer in small colloidal-sized balls, then the solution would be much less viscous. Also, if we could use surfactants to stabilise (e.g. by charging) the polymer particles in water, then there would be no need for organic solvents. Both these conditions are neatly obtained in the emulsion polymerisation process, which is schematically explained below.

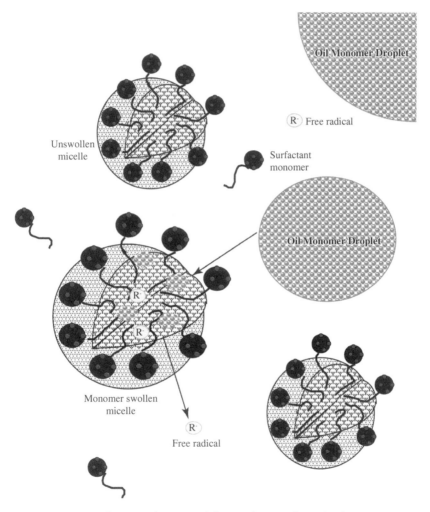

Figure 7.5   Schematic diagram of the emulsion polymerization process.

A polymer latex is produced by this process and can contain up to 50% polymer in the form of 0.1–0.5 µm spherical particles in water. A typical starting composition is:

Monomer#	50%
Soap	2%
$C_{12}H_{25}SH^*$	0.2%
$K_2S_2O_8{}^\$$	0.1%
Water	47–48%

#In commercial latex dispersions the monomer is often a
   mixture of acrylic acid and butyl and ethyl acrylates.
*A terminator or chain transfer agent to reduce the
   molecular weight of the resulting polymer.
$This is a thermal radical initiator which generates sul-
   phate radicals.

The final emulsion paint is produced by adding pigments (which are also colloidal particles), antimould agents, wetting agents and plasticizer to the latex. After spreading the paint onto a surface the water evaporates and draws the particles together until they fuse into a uniform polymer layer with embedded pigment. It should be noted that the term emulsion paint is not strictly correct because these latex paints are solid, although soft, dispersions in water, so they are colloidal solutions. This is even more the case for the original photographic emulsions.

PHOTOGRAPHIC EMULSIONS

The original photographic process was based on emulsions which were used in both black and white and colour photography, although essentially they are colloidal solutions of AgBr crystals (0.02–4 μm) dispersed in a gel of gelatin and water. On heating, the gel becomes liquid-like, and at this stage NH_4Br is added to $AgNO_3$ in the solution to give a dispersion of insoluble AgBr crystals which are fixed in the gel matrix on cooling. The emulsion is coated in a 25 μm layer on a transparent acetyl plastic base to form the film. The fundamental process on exposure to light is the production of a latent image. In this process a few surface Ag^+ ions on some of the crystals are converted to $Ag^°$ atoms. These are not large enough to see but act as nucleii for further, much more substantial conversion to $Ag^°$ atoms by a suitable reducing agent. Hence, most of the crystals with latent image nucleii are developed to give, for the case of black and white film, a black dot of colloidal dimensions. This form of chemical response to a photon signal corresponds to an amplification of about 10^6.

The detailed chemistry of photographic emulsions is very involved and encompasses solid-state physics and surface science. In colour photography, three different layers of emulsion are used where each layer is sensitized to one of the three primary colours (blue, green and red). Sensitizer molecules absorb light in each of these wavelength bands and transfer electrons to the surface of a nearby AgBr crystal. These molecules have to be within about 2 nm of an AgBr crystal to transfer the freed electron to a silver ion. During the development stage, the reducing agent itself couples to form a dye after it has been oxidised. Surface and colloid science is of fundamental importance in producing the emulsion, coating the film and controlling the dye-developer/AgBr surface reaction. In addition, finely divided colloidal TiO_2 is dispersed in printing paper to give a high level of light reflectance and brightness.

Modern photographic images are captured using a solid-state array based on a charge-coupled device (CCD) and now more commonly using the related CMOS sensors. The production of these silicon arrays is also based on an area of surface chemistry, specifically photolithography. Surface chemistry is also applied in photographic printing, especially with inkjet printers which are based on controlled droplet formation and wetting.

EMULSIONS IN FOOD SCIENCE

Colloid chemistry is very important in the production and storage of foods because of the common requirement that an organic nutrient compound must be dispersed in water (in which it may be insoluble) and this dispersion must be stabilized by surface forces. In addition, the texture or mouth feel of a food depends critically on its colloidal size distribution. A categorization of colloidal systems used in foodstuffs is given below:

Table 7.1

dispersed phase	dispersion medium	colloidal system	example
gas	liquid	foam	froth and beer
liquid	liquid	emulsion	milk, mayonnaise
solid	liquid	liquid sol	jellies, starch solution
liquid	solid	solid sol	chocolate
solid	solid	solid sol	candy

Gum arabic is a hydrocolloid (a hydrated polymer which increases viscosity) and is used to stabilise the foam on beer by reducing the rate of thinning of the soap films. The gas in the foam also seems to matter as evidenced by Guinness, whose fine cream foam is dependent on a mixture of carbon dioxide and nitrogen. Natural lipid surfactants such as the lecithins are used to stabilise oil-in-water and water-in-oil food emulsions and prevent the phases from separating out. It is for this reason that eggs (which contain lecithin) are used to mix oil and water in the production of sauces.

EMULSIONS USED FOR EXPLOSIVES IN MINING OPERATIONS

In Australia, the main use of explosives is in the initial process used in mining to break open huge areas of terrain, often whole hills, containing valuable raw minerals such as iron ore, alumina and gold. The work required is most easily supplied by simple explosive mixtures, such as ANFO prills, made from mixing ammonium nitrate and fuel oil. Similar but more waterproof AN explosives are made from water-in-oil emulsions, using concentrated ammonium nitrate solution droplets dispersed and stabilized in fuel oil by added surfactants, such as natural lecithin. A simplified schematic diagram shows the basic structure involved.

The oxidizing agent ammonium nitrate (AN) is kept separate although in close proximity by the surfactant and oil layers held between the concentrated AN cells. When the explosive composition is sensitized with gas bubbles, typically carried out on site just prior to detonation in mining processes, nitrite ions may be present due to the chemistry of the gasser solution being used, which typically contains sodium nitrite. These gas bubbles produce 'hot spots' in the material, which are important in reaction propagation via the production of local shock waves.

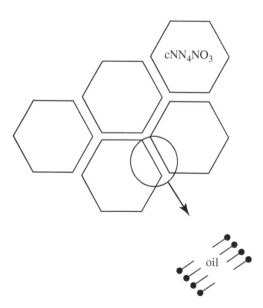

Figure 7.6 Simplified model of the type of structures formed in emulsion-based explosives.

The overall exothermic chemical reaction of ammonium nitrate and a carbon source C is:

$$2NH_4NO_3 + C = 2N_2 + 4H_2O + CO_2$$

Although ammonium nitrate prills and concentrated solutions are relatively safe, their large volume storage has led to may accidental explosions responsible for many deaths over the last 100 years or so. The exothermic thermal decomposition reaction is given by:

$$2NH_4NO_3 = 2N_2 + 4H_2O + O_2$$

Of course, it is the combination of heat production and gas production/expansion that produces the explosive effect which is of great value to the mining industry. But although undoubtedly beneficial, there are also significant dangers associated with the use of this technology.

Industrial Report

Colloid Science in Foods

Competitive adsorption of molecules at interfaces, or the displacement of one stabilising molecule by another, is an important process in food manufacture. Molecular rearrangement at interfaces not only affects the processing of foods but also the stability of the final product, the overall microstructure, and the sensory properties that the consumer ultimately experiences.

Ice cream is a complex colloidal dispersion consisting of ice particles, air bubbles, a semi-solid fat emulsion (or dispersion), protein aggregates, sugars, and polysaccharide viscosity modifiers. Ice cream is produced from a liquid mix which is emulsified (by homogenisation) to form an oil-in-water emulsion (with droplet diameter of about 0.5 to 1 µm). The protein-stabilised emulsion is then rapidly cooled so that the oil starts to crystallise and become semi-solid particles. The cooled mix is then aerated and frozen simultaneously in a high shear process. The fat dispersion in the mix prior to freezing is under constant thermal motion, and its stability to flocculation and/or coalescence depends upon the forces that act between the particles. In general, the fat droplets are stable to coalescence due to the protein that adsorbs at the oil/water interface. This leads to stabilisation through electrostatic and steric repulsion mechanisms.

Although the emulsion in the mix is relatively stable, ice cream manufacture actually requires a controlled amount of destabilisation of this emulsion. This is in order to enable fat particles to stick to the air bubbles and, to a controlled extent, one another during the aeration and freezing step. Because the fat particles are semi-solid, they form partially coalesced fat particles as opposed to fully coalescing like liquid oil-in-water emulsions, i.e. two or more partially coalesced fat droplets will retain some of their structural integrity as opposed to forming one larger spherical droplet. This process leads to several important effects: increased stability of the air bubbles to growth, slower melting of the ice cream, and a more creamy, smooth ice cream taste and texture.

Fat that is coated purely by dairy protein in the ice cream mix will remain stable even through the high shear forces of the freezer, i.e. little, or no, destabilisation will result and no fat particles will adsorb to the air. This would lead to a faster melting and less creamy ice cream product. In order to reduce the stability of the fat particles to shear in the freezing process, small molecule surface-active agents are added to the ice cream mix prior to emulsification. These typically include monoglycerides of fatty acids, or Tween. On cooling of the emulsion, they compete with the protein for space at the oil-water interface. The smaller molecules displace some of the protein, leading to an interface that is less stable to the shear and collision processes in the freezer. Therefore, they are more likely to stick to air bubbles and remain there. It is possible to add too much of the destabilising emulsifier, however. This would lead to an emulsion that is too unstable to shear. This can lead to the formation of very large fat particles through partial coalescence, a phenomenon known as "buttering". Therefore, the stability of the semi-solid oil-in-water dispersion needs to be carefully controlled in order to keep the product properties optimised.

Dr Andrew Cox
Unilever Research
Colworth
UK

Experiment

Determination of the Phase Behaviour of Microemulsions

Introduction

Micelles are formed by most surfactants (especially single-chained ones) even at fairly low concentrations in water, whereas microemulsions can be produced at much higher surfactant concentrations with, of course, the addition of an oil (e.g. decane). Microemulsions are most readily formed by double-chained surfactants. The microemulsion region for a surfactant-oil-water mixture is determined and plotted on a triangular three-phase diagram. The microemulsion region is a single, clear phase because the aggregates are too small to significantly scatter light. The double-chained cationic surfactant used in this experiment is didodecyl-dimethylammonium bromide (DDAB), and decane is used as the oil. Three-phase diagrams of this type are often measured using an aqueous salt solution as the third phase, but in this experiment we will use distilled water. An example of this type of triangular graph is given below.

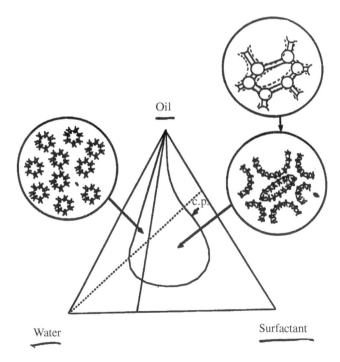

Figure 7.7   A typical three-phase triangular diagram for emulsions.

This type of graph has some interesting properties and must be used carefully. First of all, it should be noted that pure phases of the three components correspond to each apex of the triangle and that concentrations should be in either mole fractions or percentages. From this it is easy to see which concentration axis refers to any particular component. It is important to realise that the mole fraction noted on this axis relates to the line forming the base of the triangle where the apex corresponds to that of the pure phase (see diagram). Another important point is that a line drawn from any apex to the opposite axis (see diagram) corresponds to a constant ratio of mole fractions of the other two components. (In two-phase regions, tie lines have to be drawn in order to de-note the compositions of the two phases.)

Experimental procedures

The aim of the experiment is to determine the microemulsion (i.e. clear, single-phase) region for the three components already discussed and map out the results on a triangular phase diagram. The microemulsion region is determined by making up a series of mixtures in 10 cm^3 stoppered Erlenmeyer flasks with compositions that span the anticipated range. The procedure is to start with a volume of about 2 cm^3 of oil and surfactant mixtures spanning the range 20% decane to 80% decane by weight (i.e. at points along the x_o axis), weigh the sample and flask and then add increasing amounts of water up to about 70%, observing the properties of each mixture after thoroughly mixing using a vibrator table. In this way the initial ratio in the mixture of oil to surfactant remains constant, and each initial mixture runs along a different line to the pure water apex, as illustrated in the figure for the case of a starting ratio of 0.67 oil to 0.33 surfactant (by weight). At the point where a clear microemulsion phase is formed, the flask can be reweighed to obtain the amount of water added.

At low water contents (~10–20%), the mixtures will generally be milky and at some composition will become clear. At this composition a microemulsion is produced, and the boundary point has been ascertained and can be plotted. On increasing the water content, a second transition is reached (at typically about 60% water) which is more difficult to observe. This is the formation of a gel of high viscosity and marks the other boundary of the microemulsion region.

Using starting compositions of 10, 30, 50 and 70% decane, add increasing amounts of water and hence map out the microemulsion region on a triangular graph similar to that shown here (but in terms of weight percentage). Note that the initial mixtures can be made more uniform (i.e. better mixed) by adding a small amount of water (no more than 2%). Also, after each addition of water the mixture should be gently shaken and then observed before adding more. At each stage, the amount of water added must be known. Care must be taken to note down the visible properties of each mixture, such as clarity and estimated viscosity. Also, the flasks should be carefully stoppered to prevent significant loss of decane.

FOR CONSIDERATION/TYPICAL QUESTIONS

(1) What approximate composition of DDAB surfactant in water would you recommend to pump down an oil well to improve oil recovery if the main cost of the process was the cost of the surfactant?
(2) How would you expect the structure of the aggregates in the microemulsion to vary as the percentage of oil increases?

Experiment

Determination of the Phase Behaviour of Concentrated Surfactant Solutions

Introduction

Micelles are spontaneously formed by most surfactants (especially single-chained ones) even at fairly low concentrations in water, whereas at higher surfactant concentrations, with or without the addition of an oil (e.g. octane) or co-surfactant (e.g. pentanol), a diverse range of structures can be formed. These various structures include micelles, multibilayers (liquid crystals), inverted micelles, emulsions (swollen micelles) and a range of microemulsions. In each case, the self-assembled structures are determined by the relative amounts of surfactant, hydrocarbon oil, co-surfactant (e.g. pentanol) and water, as well as the fundamental requirement that there be no molecular contact between hydrocarbon and water.

In this experiment we will use various experimental techniques to attempt to identify the structures formed by a range of oil, water, surfactant and co-surfactant mixtures. The clues to this identification will come from:

(a) the composition
(b) visual observation
(c) microscopic observations with normal and polarizing light
(d) transmission of polarized light (using crossed polarizing films)
(e) viscosity
(f) electrical conductivity

The structures typically formed are listed below:
Micellar
Inverted micellar (emulsion or microemulsion)
Lamellar (liquid crystalline)
Swollen micellar (i.e. microemulsion)
Bi-continuous microemulsion
Emulsions (oil-in-water and water-in-oil)

Experimental procedures

Use the techniques listed above (a to f) to assign the structures formed by the following mixtures. You may wish to consult a demonstrator for help with identification techniques. In your report, explain the reasons for each of your structural assignments.

Make two identical surfactant solutions by dissolving 2.5 g sodium dodecyl sulfate (SDS) in 10.0 cm^3 of distilled water at room temperature. (The solution may need to be heated to dissolve the surfactant). Take care: excessive shaking will cause foaming.

To one of these solutions add pentanol sequentially, swirling to mix and observe after each mixture has equilibrated (about five minutes).

Series I

Sample No.	Amount Added/g	Total Amount Added/g
1	0	0
2	0.100	0.100
3	0.900	1.000
4*	2.250	3.250
5	5.250	8.500
6	10.250	18.750

*(**Note:** you will not be able to measure the conductivity of this sample).

Present the results in your report in the form of a table, listing the percentage composition (by weight) of each sample, the corresponding observations, and suggested structures.

After the final addition of alcohol to Series I, add 20.0 g of octane and examine the resulting mixture.

Series II

Repeat the procedure given above on the second SDS solution with octane rather than pentanol, and again construct a table of your observations and conclusions. (Note that samples 3–6 in this case are unsuitable for conductivity measurements).

FOR CONSIDERATION/TYPICAL QUESTIONS

(1) Explain why light can be transmitted through crossed polarizing films when a birefringent sample is placed between them.
(2) Why are conductivity measurements not useful for some samples?

8

Charged Colloids

The generation of colloidal charges in water. The theory of the diffuse electrical double layer. The zeta potential. The flocculation of charged colloids. The interaction between two charged surfaces in water. Laboratory project on the use of microelectrophoresis to measure the zeta potential of a colloid.

THE FORMATION OF CHARGED COLLOIDS IN WATER

Most solids release ions to some extent when immersed in a high dielectric constant liquid such as water. Even oil droplets and air bubbles have a significant surface charge in water. In order for this charging process to take place, ions must either dissociate from the surface, which will then be of opposite charge, or a specific ion could be selectively adsorbed from solution. Let us compare the work required to separate an ion pair in vacuum (or air) and in a high dielectric constant liquid, such as water. This model situation is illustrated in the following diagram.

D is the static dielectric constant of the medium, which is given by the ratio of permittivities of the medium (ε) and of free space (ε_o). From Coulomb's law for two charges, q_1 and q_2, separated by distance r, the interactive force F_c is given by the relation:

$$F_c = \frac{q_1 q_2}{4\pi D \varepsilon_o r^2} \tag{8.1}$$

Applied Colloid and Surface Chemistry, Second Edition. Richard M. Pashley and Marilyn E. Karaman.
© 2021 John Wiley & Sons Ltd. Published 2021 by John Wiley & Sons Ltd.
Companion website: www.wiley.com/go/pashley/appliedcolloid2e

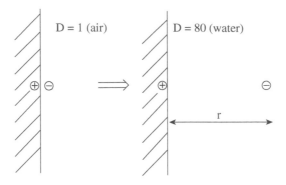

Figure 8.1   Diagram illustrating the ionisation of a surface immersed in air and in water.

where the force F_c is attractive (negative) between unlike charges. As a first approach, we can easily estimate the work (W_c) required to separate two charges from close separation of, say, 2 Å to a large distance by integration of Equation (8.1). In water, where $D = 80$, this work is about 6×10^{-21} J, which is quite close to the kinetic energy of the free ion (i.e. kT). However, if the medium was air or vacuum ($D = 1$), the work required would be about 100 kT. Clearly, it is the high dielectric constant or polar nature of water which allows this ion dissociation to occur, whereas in air and non-polar liquids (e.g. $D \sim 2$ for liquid hexane) we would expect no dissociation. It is for this reason that we are mostly interested in charging processes at the solid/aqueous interface and the stability of colloids, which often become charged when dispersed in aqueous media.

   Although the single ion dissociation approach gives a good indication of the basic conditions for dissociation, the real situation is more complicated because there will usually be a high density of ions dissociated from the surface. This will introduce repulsive forces between the dissociated ions and a much stronger attraction back to the surface, because of the high electric field generated there. In fact, the electric field generated at the surface prevents the dissociated ions from leaving the surface region completely, and these ions, together with the charged surface, form a diffuse electrical double layer, as illustrated in the following figure.

   In order to understand the diffuse layer in detail, we need to go back to the fundamental equations of electrostatics due to J. C. Maxwell. The equation of interest relates the local electric field $\vec{E}(\vec{r})$ at the position vector $\vec{r}$ to the net local electric charge density $\rho(\vec{r})$:

$$\vec{\nabla} \bullet \vec{E}(\vec{r}) = \frac{\rho(\vec{r})}{\varepsilon_0 D} \tag{8.2}$$

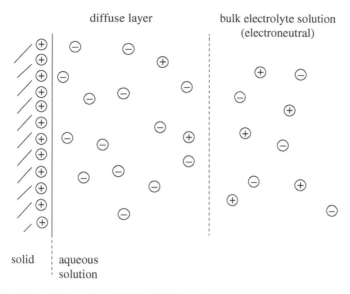

Figure 8.2 The diffuse electrical double layer in aqueous solution next to a flat charged surface.

Thus, the divergence or flux of the electric field is directly related to the net charge at that point. The electric field is simply defined as the force acting on a unit charge:

$$\vec{E}(\vec{r}) = \lim_{q \to 0} \frac{\vec{F}(\vec{r})}{q} \qquad (8.3)$$

The limit is required because otherwise a finite test charge q would itself perturb the electric field. Clearly from equation (8.3), the direction of the electric field is that taken by a positive charge when free to move. Because the electric field is a vector, it is often more useful to use the corresponding electrostatic potential $\psi(\vec{r})$, which is a scalar quantity and is defined as the potential energy gained by moving a unit charge from infinity to the position $\vec{r}$. The potential energy of an ion of charge $Z_i q$ at $\vec{r}$ is therefore simply given by $Z_i q \psi(\vec{r})$, where q is the proton charge and Z_i the valency of the ion. Note that if the local potential is positive, then a positive charge will have an increased (positive) potential energy at that position compared with at infinity.

Since $\psi(\vec{r})$ is the electrostatic potential energy per unit charge, the gradient of this parameter with distance must be equal to the force acting on a unit charge – which is the definition of the electric field. Hence it follows that:

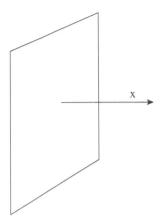

Figure 8.3   The one-dimensional case of a flat surface.

$$\vec{E}(\vec{r}) = -\nabla \psi (\vec{r}) \tag{8.4}$$

In order to simplify equation (8.2), let us consider the common case of the electric field generated by a charged flat surface (such as an electrode):
For this case and using electric potentials, equation (8.2) becomes:

$$\frac{d^2\psi(x)}{dx^2} = -\frac{\rho(x)}{\varepsilon_o D} \tag{8.5}$$

where ψ and ρ are now only functions of x, the distance from the flat surface. To solve this equation, we need to find the relationship between $\rho(x)$ and $\psi(x)$. The local density of any ion of charge Z_iq (which can be either positive or negative) must depend on its electrostatic potential energy at that position (i.e. at x). From our definition of ψ, this potential energy is given by $Z_iq\psi(x)$. Since any ion next to an immersed charged surface must be in equilibrium with the corresponding ions in the bulk solution, it follows that the electrochemical potential of an ion at distance x from the surface must be equal to its bulk value. Thus:

$$\mu_i^b = \mu_i^x = \mu_i^0 + kT \ln C_i(B) = \mu_i^0 + Z_iq\psi(x) + kT \ln C_i(x)$$

where $C_i(B)$ and $C_i(x)$ are the ion concentrations in bulk and at distance x from the charged surface, and it is assumed that these are dilute solutions (i.e. $\psi(B) = 0$). This equation leads directly to the important Boltzmann distribution, which can be used to obtain the concentration at any other electrostatic potential energy by the familiar relationship:

$$C_i(x) = C_i(B)exp\left[-\frac{q\psi(x)Z_i}{kT}\right] \qquad (8.6)$$

This result is very useful because it gives us the concentration of any ion next to a charged surface when immersed in an electrolyte solution.

We can also use this result to obtain the net charge density (due to both counterions and co-ions) at a distance x from a charged surface, since:

$$\rho(x) = \sum_i Z_i q C_i(x)$$

$$= \sum_i Z_i q C_i(B)exp\left[-\frac{q\psi(x)Z_i}{kT}\right] \qquad (8.7)$$

and substitution in equation (8.5) then gives the result:

$$\frac{d^2\psi(x)}{dx^2} = -\frac{q}{\varepsilon_o D}\sum_i Z_i C_i(B)exp\left[-\frac{q\psi(x)Z_i}{kT}\right] \qquad (8.8)$$

This is the important Poisson-Boltzmann (P-B) equation, and the model used to derive it is usually called the Gouy-Chapman (G-C) theory. It is the basic equation for calculating all electrical double-layer problems for flat surfaces immersed in electrolyte solutions. In deriving it we have, however, assumed that all ions are point charges and that the potentials at each plane x are uniformly smeared out along that plane. These are usually reasonable assumptions, especially for dilute solutions.

We can now set about using equation (8.8) to give us information about the quantitative details of the electrical double layer. Let us, for simplicity, assume that the electrolyte in which the surface is immersed is symmetrical, that is a Z:Z electrolyte (i.e. 1:1, 2:2 or 3:3, where $Z = 1, 2$ or 3). Equation (8.7) then becomes:

$$\rho(x) = ZqC(B)\left[exp\left[-\frac{q\psi(x)Z}{kT}\right] - exp\left[\frac{q\psi(x)Z}{kT}\right]\right]$$

and if we, for convenience, define $Y = Zq\psi(x)/kT$ this reduces to:

$$\rho(x) = -ZqC(B)\{exp(Y) - exp(-Y)\}$$

$$\therefore \rho(x) = -2ZqC(B)sinh(Y) \qquad (8.9)$$

which in the P-B equation (8) gives:

$$\frac{d^2\psi(x)}{dx^2} = \frac{2Z^2q^2C(B)}{\varepsilon_o DkT}sinh(Y) \tag{8.10}$$

where for symmetrical electrolytes: $C(B) = C_i(B)$.

Now, for convenience (and not arbitrarily, as will be seen later), let us replace the real distance x with a scaled, dimensionless distance X, such that $X = \kappa x$ and κ^{-1}, which is called the Debye length, is defined as:

$$\kappa^{-1} = \left[\frac{\varepsilon_o DkT}{q^2 \sum_i C_i(B)Z_i^2}\right]^{1/2} \tag{8.11}$$

The parameter κ^{-1} turns out to be very useful, has the units of length and depends on both the electrolyte type and concentration. Scaling x in equation (8.10) thus gives the deceptively simple non-linear second-order differential equation:

$$\frac{d^2Y}{dX^2} = sinh\,Y \tag{8.12}$$

Integration of equation (8.12) gives the potential distribution next to a charged (flat) surface:

$$Y = 2\,ln\left[\frac{1+\gamma\exp(-X)}{1-\gamma\exp(-X)}\right] \tag{8.13}$$

where

$$\gamma = \left[\frac{\exp\left(\frac{Y_0}{2}\right)-1}{\exp\left(\frac{Y_o}{2}\right)+1}\right]$$

and Y_o is the scaled electrostatic potential at the surface of the charged plane (i.e. at $x = 0$, $\psi = \psi_o$ and at $X = 0$, $Y = Y_o$). This result gives us a great deal of information about the decay in potential and hence the distribution of all ionic species in the diffuse electrical double layer next to a charged flat surface immersed in an electrolyte solution.

THE DEBYE LENGTH

We can easily get some idea of what this theory predicts by looking at the limit of low potentials, that is where $Y_o \ll 1$ (which for 1:1 electrolytes corresponds to $\psi_o < 25$ mV). In this case equation (8.13) can be shown to reduce to the simple result:

$$\psi(x) \cong \psi_o exp(-\kappa x) \qquad (8.14)$$

which demonstrates the physical meaning of the Debye length. This length is also referred to as the 'double-layer thickness', and from equation (8.14), it obviously gives an indication of the extent of the diffuse layer into the solution. It is the distance from the surface where the surface potential has fallen to 1/e of its original value. Equation (8.14), although approximate, enables us to estimate the decay in electrostatic potential away from a flat charged surface. Some typical results are shown in the following diagram.

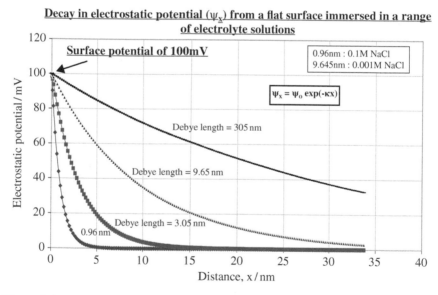

Figure 8.4   Estimates of the decay in electrostatic potential away from a charged flat plate in a range of electrolyte solutions.

These results clearly demonstrate the effect of electrolyte concentration on screening the range of the surface electrostatic potential. In each case the

Debye length is equal to the distance from the surface at which the surface potential has fallen to 100/e or 37 mV. (Note that use of Equation (8.13) above gives a more accurate potential distribution.)

Once the electrostatic potential distribution next to a charged surface is known, the Boltzmann distribution equation (8.6) can be used to calculate the corresponding distribution of both counter-ions and co-ions in an electrolyte solution next to a charged surface. The results obtained for both counter-ions and co-ions is given in the diagrams below for a monovalent electrolyte solution, e.g. NaCl electrolyte at 1 mM concentration, surrounding a +100 mV flat surface.

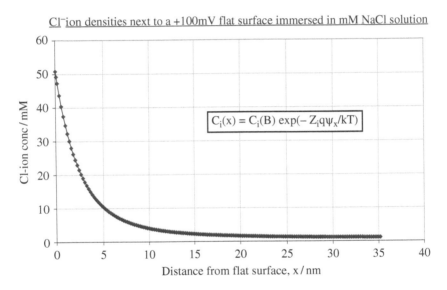

Figure 8.5 Estimates of the Cl⁻ counter-ion concentration away from a flat charged surface of potential +100 mV, when immersed in a mM NaCl solution.

Using the relatively simple equations derived above, it is clearly possible to obtain precise details about the ion distributions next to charged flat surfaces when they are immersed in electrolyte solutions. This information is crucial to a thorough understanding of the properties of electrodes, colloidal solutions and many areas of industrial chemistry. As might be expected, counter-ions are pulled towards surfaces of opposite charge, reaching quite high concentrations close to the surface. By comparison, co-ions are expelled by the same-charge surface and form a depleted layer. Similar results can be obtained using a combination of mixed and multivalent electrolytes.

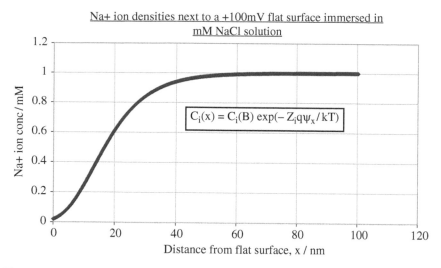

Figure 8.6   Estimates of the Na+ co-ion concentration away from a flat charged surface of potential +100 mV, when immersed in a mM NaCl solution.

THE SURFACE CHARGE DENSITY

So far, we have talked mostly in terms of electrostatic potentials. But can we use this theory to find the corresponding charge density on a surface (σ_0)? In order for the electrical double layer to be neutral overall, it follows that the total summed charge in the diffuse layer (i.e. in solution) must be equal to the surface charge. Thus, it follows that:

$$\sigma_D = -\sigma_0 = \int_0^\infty \rho(x)\,dx \qquad (8.15)$$

where σ_D is the total diffuse layer surface charge density. Now since from equation (8.5):

$$\frac{d^2\psi(x)}{dx^2} = \frac{-\rho(x)}{\varepsilon_o D} \qquad (8.5)$$

it follows that:

$$\sigma_D = \varepsilon_0 D \int_0^\infty \frac{d^2\psi(x)}{dx^2}\,dx$$

and hence

$$\sigma_0 = -\varepsilon_0 D \left[\frac{d\psi(x)}{dx} \right]_{x=0} \tag{8.16}$$

since $d\psi(x)/dx = 0$ as $x \to \infty$.

Now we can integrate equation (8.8) using the fact that

$$\int \frac{d^2y}{dx^2} dy = 1/2 \left(\frac{dy}{dx} \right)^2 + C \tag{8.17}$$

and using various known boundary conditions obtain the first order differential equation:

$$\frac{d\psi(x)}{dx} = - \left[\frac{2kT}{\varepsilon_0 D} \sum_i C_i (B) \left(exp \left[-\frac{Z_i q\psi(x)}{kT} \right] - 1 \right) \right]^{1/2} \tag{8.18}$$

which on substitution into equation (8.16) gives the result:

$$\sigma_0 = (sgn\psi_0) \left[2\varepsilon_0 DkT \sum_i C_i (B) \left(exp \left[-\frac{Z_i q\psi_0}{kT} \right] - 1 \right) \right]^{1/2} \tag{8.19}$$

which, for a 1:1 electrolyte reduces to:

$$\sigma_o = \left[8\varepsilon_o DkTC(B) \right]^{\frac{1}{2}} sinh \left(\frac{q\psi_0}{2kT} \right). \tag{8.20}$$

Hence, the surface charge density (in C/m²) can be easily calculated from the surface potential (in V) for a planar surface immersed in an electrolyte solution of known concentration.

THE ZETA POTENTIAL

The stability of many colloidal solutions depends critically on the magnitude of the electrostatic potential (ψ_0) at the surface of the colloidal particles. One of the most important tasks in colloid science is therefore to obtain an estimate of ψ_0 under a wide range of electrolyte conditions. In practice, one of the most convenient methods for obtaining ψ_0 uses the fact that a charged particle will move at some constant, limiting velocity under the influence of an applied electric field. Even quite small particles (i.e. < 1 μm) can be observed using a dark-field microscope and their velocity directly measured. This technique is called microelectrophoresis, and what is measured is the electromobility (μ) of a colloid, which is its speed (u) divided by the applied electric field (E).

Let us now examine how we can obtain an estimate of ψ_0 from the measured electromobility of a colloidal particle. It turns out that we can obtain simple analytic equations only for the cases of very large and very small particles. Thus, if a is the radius of an assumed spherical colloidal particle, then we can obtain direct relationships between electromobility and the surface potential, if either κa >100 or κa < 0.1, where κ^{-1} is the Debye length of the electrolyte solution. Let us first look at the case of small spheres (where κa < 0.1), which leads to the Huckel equation.

THE HUCKEL EQUATION ($\kappa a < 0.1$)

The spherically symmetric potential around a small charged sphere is described by the Poisson-Boltzmann equation (compare Equation (8.2)) for this geometry:

$$\nabla^2 \psi (r) = -\frac{\rho(r)}{\varepsilon_0 D} \qquad (8.21)$$

where $\rho(r)$ is the average charge density and $\psi(r)$ the potential at a distance r away from the central charge. This equation can be simplified using the Debye-Huckel or linear-approximation, which is valid for low potentials:

$$\nabla^2 \psi (r) = \kappa^2 \psi (r) \qquad (8.22)$$

which has the simple general solution:

$$\psi = \frac{A\exp(\kappa r)}{r} + \frac{B\exp(-\kappa r)}{r} \qquad (8.23)$$

The constant A must equal zero for the potential ψ to fall to zero at a large distance away from the charge, and the constant B can be obtained using the second boundary condition, that: $\psi = \psi_0$ at $r = a$, where a is the radius of the charged particle and ψ_0 the electrostatic potential on the particle surface. Thus, we obtain the result that:

$$\psi_0 = \frac{B\exp(-\kappa r)}{r} \tag{8.24}$$

and, therefore

$$\psi = \frac{\psi_0 a \exp\left[-\kappa(r-a)\right]}{r} \tag{8.25}$$

The relationship between the total charge q on the particle and the surface potential is obtained using the fact that the total charge in the electrical double layer around the particle must be equal and of opposite sign to the particle charge, that is:

$$q = -\int_a^\infty 4\pi r^2\, \rho(r)\,dr \tag{8.26}$$

where $\rho(r)$ is the charge density at a distance r from the centre of the charged particle. The value of $\rho(r)$ can be obtained from combination of Equations (8.1) and (8.2), assuming the linear approximation is valid and hence:

$$q = 4\pi\varepsilon_0 D\kappa^2 \int_a^\infty r^2 \psi\, dr \tag{8.27}$$

Now, using equation (8.25) for ψ:

$$q = 4\pi\varepsilon_0 D\kappa^2 a\psi_0 \int_a^\infty r \exp\left[-\kappa(r-a)\right]dr \tag{8.28}$$

Integration using Leibnitz's theorem gives:

$$q = 4\pi\varepsilon_0 Da\psi_0 \left(1+\kappa a\right) \tag{8.29}$$

and rearranging this equation leads to a useful physical picture of the potential around a sphere, thus:

$$\psi_0 = \frac{q}{4\pi D\varepsilon_0 a} - \frac{q}{4\pi D\varepsilon_0 \left(a + \kappa^{-1}\right)} \tag{8.30}$$

This result corresponds to a model of the charged particle with a diffuse layer charge (of opposite sign) at a separation of $1/\kappa$, as illustrated below:

Since we now have an equation (8.29) which relates the charge on the particle to the surface potential, we can combine this with the forces acting on a moving particle in an applied electric field. Thus when the particle is moving at

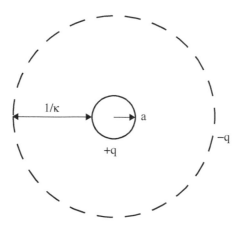

Figure 8.7 Simple model of the diffuse, averaged electrical double layer around a small charged colloid.

constant velocity (u), the electrostatic force on the particle (qE) must equal the drag force, which we may assume (for laminar, steady fluid flow) to be that given by Stokes's law (i.e. $F_{drag} = 6\pi a u \eta$). Using equation (8.29) and the fact that we define the electromobility (μ) of a particle as u/E, we obtain the result that:

$$\psi_0 = \frac{3\mu\eta}{2\varepsilon_0 D(1+\kappa a)}$$

which for $\kappa a \ll 1$ becomes:

$$\psi_0 = \frac{3\mu\eta}{2\varepsilon_0 D} = \varsigma$$

(8.31)

In this result, the condition of small particles means that the actual size of the particles (which is often difficult to obtain) is not required. For reasons to be discussed later, we will call the potential obtained by this method the zeta potential (ζ) rather than the surface potential. In the following section we consider the alternative case of large colloidal particles, which leads to the Smoluchowski equation.

THE SMOLUCHOWSKI EQUATION ($\kappa a > 100$)

Let us now consider an alternative derivation for the case of large colloidal particles, where the particle radius is much larger than the Debye length (i.e. $\kappa a > 100$). The situation is best described by the following schematic diagram, where the surface of the large particle is assumed to be effectively flat relative to the double layer thickness. It is also assumed, in this approach, that the fluid flows past the surface of the particle in parallel layers of increasing velocity, with distance from the surface. At the surface, the fluid has zero velocity (relative to the particle), and at a large distance away, the fluid moves with the same velocity as the particle but in the opposite direction. It is also assumed that the flow of the fluid does not alter the ion distribution in the diffuse double layer (i.e. in the x direction).

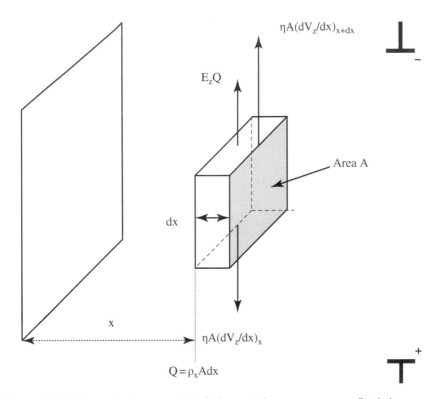

Figure 8.8 Schematic diagram of the balance in forces acting on a fluid element close to the surface of a large colloidal particle.

Under these conditions, mechanical equilibrium can be considered in a fluid element between x and $x + dx$, when the viscous forces acting in the z direction on the fluid element, due to the velocity gradient in the x direction, are precisely balanced by the electrostatic body force acting on the fluid due the charge contained in it. Thus, we obtain the mechanical equilibrium condition that:

$$E_z \rho_x A dx = \eta A \left(\frac{dV_z}{dx} \right)_x - \eta A \left(\frac{dV_z}{dx} \right)_{x+dx} \tag{8.32}$$

or

$$E_z \rho_x A dx = -\eta A \left(\frac{d^2 V_z}{dx^2} \right) dx \tag{8.33}$$

We can then relate the charge density, ρ_x, to the electrostatic potential using the one-dimensional Poisson-Boltzmann equation:

$$\frac{d^2 \psi}{dx^2} = -\frac{\rho_x}{\varepsilon_0 D} \tag{8.34}$$

Thus, in Equation. 8.34:

$$E_z \varepsilon_0 D \frac{d^2 \psi}{dx^2} dx = \eta \left(\frac{d^2 V_z}{dx^2} \right) dx \tag{8.35}$$

which on integration gives:

$$E_z \varepsilon_0 D \frac{d\psi}{dx} = \eta \left(\frac{dV_z}{dx} \right) + c_1 \tag{8.36}$$

Since $d\psi/dx = 0$ when $dVz/dx = 0$, the integration constant, c_1, must be equal to zero and a second integration:

$$\int_{\psi=0}^{\psi=\varsigma} E_z \varepsilon_0 D \frac{d\psi}{dx} dx = \int_{Vz}^{0} \eta \left(\frac{dV_z}{dx} \right) dx \tag{8.37}$$

produces the result that:

$$E_z \varepsilon_0 D \varsigma = -\eta V_z \tag{8.38}$$

if it is assumed that $D \neq f(x)$ and $\eta \neq f(x)$ (i.e. that the fluid is Newtonian).

Since $-V_z$ refers to the fluid velocity, we can easily convert Equation (8.38) to particle velocity (i.e. $V_p = -V_z$), and from our definition of electromobility (μ), it follows that:

$$\varsigma = \frac{\mu \eta}{\varepsilon_0 D} \tag{8.39}$$

This important result is called the Smoluchowski equation and, as before, the zeta potential is directly related to the mobility and does not depend on either the size of the particle or the electrolyte concentration.

In summary, for the two extreme cases, we have:

$$\varsigma = \frac{3\mu\eta}{2\varepsilon_0 D}, \text{ for } \kappa a \ll 1 \ (<0.1)$$

$$\varsigma = \frac{\mu\eta}{\varepsilon_0 D}, \text{ for } \kappa a \gg 1 \ (>100)$$

CORRECTIONS TO THE SMOLUCHOWSKI EQUATION

These equations have been shown to be correct under the conditions of electrolyte concentration and particle size stated. However, it is easy to show that using typical colloidal sizes and salt concentrations, many colloidal systems of interest, unfortunately, will fall between the ranges covered by these equations. In deriving both the Huckel and Smoluchowski equations we have, for simplicity, ignored relaxation and electrophoretic retardation effects, which have to be included in a more complete theory. The origin of these effects is illustrated in the following diagram of a solid charged particle moving in an electric field.

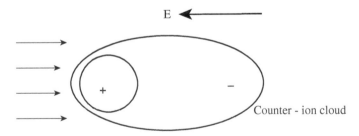

Figure 8.9 Diagram of a charged colloid moving in a fluid under the action of an applied electric field.

When in motion, the diffuse electrical double layer around the particle is no longer symmetrical, and this causes a reduction in the speed of the particle compared with that of an imaginary charged particle with no double layer. This reduction in speed is caused by both the electric dipole field set up which acts in opposition to the applied field (the relaxation effect) and an increased viscous drag due to the motion of the ions in the double layer which drag liquid with them (the electrophoretic retardation effect). The resulting combination of electrostatic and hydrodynamic forces lead to rather complicated equations which, until recently, could only be solved approximately. In 1978, White and O'Brien (*J. Chem. Soc.*, Faraday II, 741, 1607) developed a clever method of numerical solution and obtained detailed curves over the full range of κa values ($0 \to \infty$) and surface potentials. At intermediate values of κa, the relationship between ζ and μ is non-linear and strongly dependent on electrolyte type (i.e. charge and ion diffusion coefficients). A computer program is available from the authors to enable calculation of the zeta potential for any common electrolyte. With the successful introduction of this precise numerical procedure, the onus is now on experimentalists to carry out well-defined mobility measurements on ideal spherical particles of accurately known size.

The following diagram shows the non-dimensional electrophoretic mobility as a function of zeta potential for a range of κa values, corrected for these two factors. The curves represent computed values obtained by White and O'Brien, while the broken lines are the thin double-layer approximation. The $\kappa a = \infty$ line is Smoluchowski's result. (See also the standard text: R. J. Hunter, *Zeta Potential in Colloid Science,* London: Academic Press, 1981.)

Finally, let us return to the problem of relating the measured zeta potential to the defined surface potential. The zeta potential is always measured (by definition) in an electrokinetic experiment. In this case, the fluid has to flow around the particle. We expect, however, that a certain thickness of fluid (of roughly molecular dimensions) will remain stationary with respect to the particle, due to the large amount of work required to move fluid molecules along a solid surface. Obviously, there will not be a sharp cut-off at say one or two molecular layers, but a gradual increase in fluid flow will occur away from the particle. Since the total charge on the solid particle is responsible for the surface potential, the measured value, zeta, is generally of slightly lower magnitude. As we will see later, a better estimate of the surface potential can be obtained from direct interaction force measurements, and values so obtained can be compared with electrokinetic measurements, on exactly the same system in some cases (such as for muscovite mica). Some systems do show excellent agreement between zeta and the surface potential, whilst others differ significantly. Since no definitive results have yet been obtained, it is perhaps best to assume as a first approximation that these two potentials are similar.

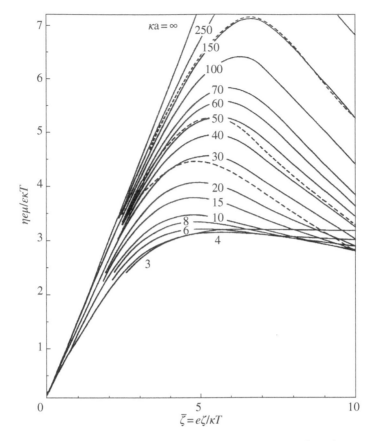

Figure 8.10 Theoretical calculations of the corrections required to obtain zeta potentials (ζ) from measured electromobilities (μ).

THE ZETA POTENTIAL AND FLOCCULATION

A good example of the use of microelectrophoresis experiments is supplied by the study of ferric flocs, which are widely used in municipal water treatment plants. The zeta potentials shown below were derived from the measured floc electromobilities using the Smoluchowski equation. The figure below shows the effect of pH on the zeta potential of ferric flocs generated by precipitation of 3.3 mg/L (20 µM) and 33 mg/L (200 µM) ferric chloride in mM sodium chloride solution. Since almost all colloidal contaminants in natural water systems are negatively charged, it is important to operate under pH conditions where the flocs are positively charged, i.e. at pH values above the isoelectric point (i.e.p.), where the particles are uncharged. Under these conditions, the flocs have an electrostatic force attracting the contaminant particles to the flocs. The large flocs, containing adsorbed contaminants, are then sedimented and collected via filtration through high flow-rate sand bed filters. Unfortunately, in practice ferric chloride usually contains manganese as a contaminant, and this necessitates precipitation at pH values between 8 and 9 to prevent manganese dissolution, which will colour the drinking water. The use of these higher pH values means that polycationic water-soluble polymers are often added, which adsorb on to the ferric flocs and increase their i.e.p.

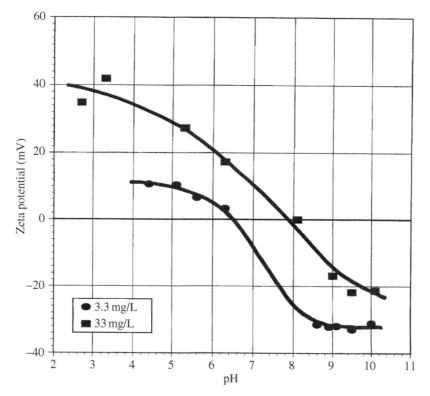

Figure 8.11 Measured zeta potentials of ferric flocs as a function of concentration and pH.

THE INTERACTION BETWEEN ELECTRICAL DOUBLE LAYERS

So far, we have used the Maxwell equations of electrostatics to determine the distribution of ions in solution around an isolated charged flat surface. This distribution must be the equilibrium one. Hence, when a second surface, also similarly charged, is brought close, the two surfaces will 'see' each other as soon as their diffuse double layers overlap. The ion densities around each surface will then be altered from their equilibrium value, and this will lead to an increase in energy and a repulsive force between the surfaces. This situation is illustrated schematically below for non-interacting and interacting flat surfaces:

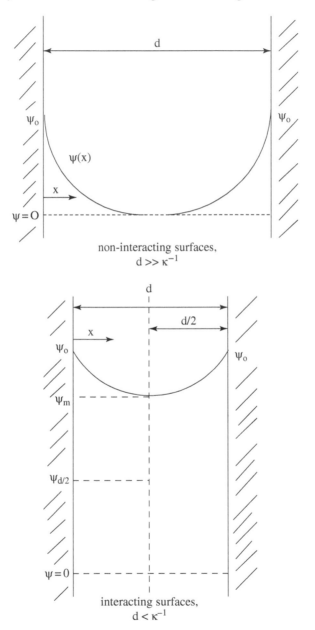

Figure 8.12 Schematic diagram of non-interacting and interacting charged surfaces immersed in an electrolyte solution.

The repulsive electrostatic forces generated between similarly charged particles will act to prevent coagulation and stabilise colloidal solutions. If this repulsion were absent (for example, by neutralisation of the surface charge), attractive van der Waals forces (to be discussed later) would cause each particle-particle collision to be successful. That is, particles colliding due to random kinetic motion in the solution would adhere to each other, forming large aggregates which would settle out from solution. The system would coagulate. We can now apply the Poisson-Boltzmann model to this important interaction.

If we consider the interaction of two identical planar surfaces (see the last schematic figure), we see that at the mid-plane there is no net electric field, that is at $x = d/2$, $d\psi(x)/dx = 0$ and $\psi(x) = \psi_m$. Thus, even though there is a net (non-zero) charge on the fluid at the mid-plane, no net electrostatic force acts upon it. Now, since the mid-plane potential is not zero for interacting surfaces, there must be a higher concentration of ions at this plane compared with in the bulk solution, from the Boltzmann distribution. This higher concentration, at the mid-plane, will give rise to a higher osmotic pressure at this plane, relative to the bulk solution, which will push the surfaces apart by drawing in more solvent (water). Since this pressure must be the same throughout the liquid film (between the surfaces) and because only osmotic forces act at the mid-plane, the pressure acting between the surfaces must be equal to the increased osmotic pressure at the mid-plane over that in bulk. This osmotic pressure difference can be calculated simply by determining the total concentration of ions at the mid-plane (C). This concentration can be easily calculated using the Boltzmann distribution equation, if the mid-plane potential, ψ_m, is known, thus:

$$C_m^T = \sum_i C_i(B) exp\left[\frac{Z_i q\psi_m}{kT}\right] \tag{8.40}$$

which for (symmetrical) $Z:Z$ electrolytes becomes:

$$C_m^T = C(B)\left[exp\left[-\frac{Zq\psi_m}{kT}\right] + exp\left[\frac{Zq\psi_m}{kT}\right]\right] \tag{8.41}$$

or

$$C_m^T = 2C(B)cosh\left[\frac{Zq\psi_m}{kT}\right] \tag{8.42}$$

For ideal solutions, the osmotic pressure is simply given by CkT. Hence, the osmotic pressure difference between the mid-plane region and the bulk solution, which must be equal to the repulsive pressure in the film Π_R, is given by the equation:

$$\Delta\Pi_{os} = \Pi_R = kT\left[C_m^T - 2C(B)\right]$$

and hence:

$$\Pi_R = 2C(B)kT\left[\cosh\left(\frac{Zq\psi_m}{kT}\right) - 1\right] \tag{8.43}$$

If we can now determine ψ_m as a function of the separation distance d between the surfaces, we can calculate the total double-layer (pressure) interaction between the planar surfaces. Unfortunately, the P-B equation cannot be solved analytically to give this result, and instead numerical methods have to be used. Several approximate analytical equations can, however, be derived, and these can be quite useful when the particular limitations chosen can be applied to the real situation.

One of the simplest equations is obtained using the Debye-Hückel approximation (for low potentials) and the superposition principle. The latter assumes that the unperturbed potential near a charged surface can be simply added to that potential due to the other (unperturbed) surface. Thus, for the example shown in the last figure, it follows that $\psi_m = 2\psi_{d/2}$. This is precisely valid for Coulomb-type interactions, where the potential at any point can be calculated from the potentials produced by each fixed charge individually. However, the Poisson-Boltzmann equation is non-linear (this has to do with the fact that in the diffuse double layer the ions are not fixed but move because of their kinetic energy) and so this is formally not correct but it still offers a useful approximation.

Using the Debye-Hückel (D-H) approximation, the potential decay away from each flat surface is given by:

$$\psi(x) \approx \psi_o \, exp(-\kappa x)$$

and hence using the superposition principle:

$$\psi_m \approx 2\psi_{d/2} \approx 2\psi_o \, exp(-\kappa d / 2) \tag{8.44}$$

Again, if we use the D-H approximation in equation (8.44), we can expand the cosh function for small values of x (or ψ_0) to give:

$$\cosh x = \left(1 + x^2/2\right) or \cosh\left(Zq\psi_m/kT\right) = 1 + \frac{1}{2}\left(zq\psi_m/kT\right)^2$$

which simplifies to:

$$\Pi_R \sim C(B)kT \times \left(\frac{Zq\psi_m}{kT}\right)^2 \tag{8.45}$$

Hence, from equations (8.44) and (8.45):

$$\Pi_R \approx \frac{C(B)Z^2q^2}{kT} 4\psi_o^2 exp(-\kappa d)$$

which on using the definition of the Debye length (κ^{-1}) becomes:

$$\Pi_R \approx 2\varepsilon_o D\kappa^2 \psi_o^2 exp(-\kappa d) \tag{8.46}$$

This result shows us that the repulsive double-layer pressure (for the case of low potentials) decays exponentially with a decay length equal to the Debye length and has a magnitude which depends strongly on the surface potential.

The corresponding interaction energy V_F between flat surfaces can be obtained by integrating the pressure from large separations down to separation distance, d:

$$V_F = -\int_0^\infty \Pi_R d(d) \tag{8.47}$$

which gives:

$$V_F \approx 2\varepsilon_o D\kappa \psi_o^2 exp(-\kappa d) \tag{8.48}$$

So far, we have only considered the interaction between flat surfaces, basically because of the simplification of the P-B equation in one dimension. Of course, colloidal particles are usually spherical, and for this geometry the exact numerical solution of the three-dimensional P-B equation becomes very difficult. However, we can obtain an estimate of the sphere-sphere interaction from the planar result if the radius a of the spheres is much larger than the Debye length (i.e. $\kappa a \gg 1$). The method was developed by Boris Derjaguin and is considered in the next section.

THE DERJAGUIN APPROXIMATION

A schematic diagram of the analysis method used is given in the following figure.

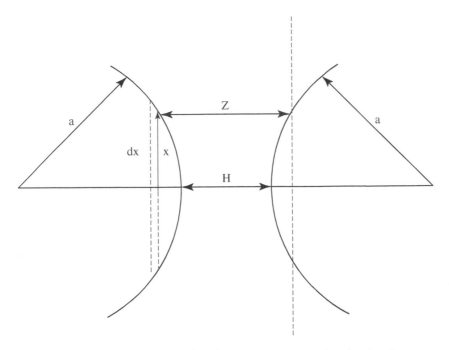

Figure 8.13 Diagram used to explain the Derjaguin approximation for the interaction between two spheres.

In this procedure, called the Derjaguin approximation, we consider the interaction of the circular annulus (dx) with an imaginary parallel surface plane at distance Z. With this assumption, the total interaction energy V_s between the spheres is then given by:

$$V_S \approx 2\pi \int_{Z=H}^{Z=\infty} V_F x\,dx \qquad (8.49)$$

From simple geometry we can easily rearrange this equation in the form:

$$V_S \approx \pi a \int_{Z=H}^{Z=\infty} V_F\,dZ \qquad (8.50)$$

Using the result given in equation (8.48), we can then obtain the corresponding interaction energy between spheres:

$$V_S = 2\pi a \varepsilon_o D\psi_o{}^2 exp(-\kappa H) \qquad (8.51)$$

Once again, the interaction energy decays exponentially with separation distance and is strongly dependent on both the surface potential and the electrolyte concentration.

Industrial Report

The use of emulsions in coatings

From 2000 BC, when decorative paintings adorned the walls of Egyptian tombs, all the way up through much of the 1900s, little changed in the rudimentary approach to formulating decorative or protective paints. Paints were based on naturally occurring oleoresinous materials: combinations of naturally occurring drying oils and resins. Linseed oil became the most widely used oil, while amber was the most common resin. Eventually, naturally occurring resins were replaced by synthetic resins, such as alkyds. The resins and drying oils perform the role of the 'binder' in the paint; that is, they bind the pigment particles and stick to the substrate being painted – they are the 'glue'. These types of paints were usually thinned with organic solvents; consequently they are flammable and often represent a health hazard. The high levels of organic solvents present in these coatings causes considerable pollution concerns.

The architectural coatings industry was revolutionized around 1950, when coatings based on waterborne emulsions were invented. This paradigm shift represents the most significant invention in the history of coatings, due to the immeasurable benefits in odour, toxicity, flammability, ease of handling, clean-up and often performance of emulsions, compared to solvent/oil-based products. In the new water-based coatings, it is the microscopic emulsion particles which coalesce, act as the binder that holds the pigment particles together, form a continuous film and are the glue, sticking to the substrate. Emulsion particles are essentially tiny plastic particles which are dispersed in water.

Coatings emulsions are generally formed by addition polymerization of common, highly available monomers, using free radical initiators to create polymers having molecular weights from a few thousand up to millions. The polymerization is most often stabilized by non-ionic and/or anionic surfactants, which emulsify the insoluble monomer droplets and then stabilize the resulting particles, usually in the shape of a sphere. In addition to surfactants, emulsions are sometimes stabilized with water soluble polymers, which act as resin support for the growing polymer particles. Also, many coatings emulsion polymers contain ionic groups which enhance stability via contributing to an electrical double layer.

Commercially significant coatings emulsions include acrylics; copolymers of acrylates, such as butyl acrylate, and methacrylates, such as methyl methacrylate; styrene-acrylics; copolymers of styrene with an acrylate monomer; and vinyl acetate polymers: homopolymers of vinyl acetate or copolymers with softer monomer such as ethylene or butyl acrylate.

Polymers which are 100% acrylic are known for their outstanding exterior durability properties, as well as excellent alkali resistance and overall high performance. Styrene is generally lower cost than many other monomers, so styrene-acrylics are lower in cost than all acrylics but have poorer exterior durability because styrene is a UV absorber and degrades. Styrene-acrylics do, however, generally have very good water resistance properties due to their hydrophobicity. Styrene acrylics are popular in some areas of Europe, Asia and Latin America. Vinyl acetate is also somewhat lower in cost, so these polymers are popular for interior paints, particularly in the United States, where exterior durability and alkali resistance are not performance issues.

Most emulsion polymers are spheres, generally the lowest energy and therefore most stable configuration. However, there are other particle shapes and morphologies which can be obtained during emulsion polymerization, when special properties are desired which can be achieved via a unique morphology. Several alternative morphologies are shown below.

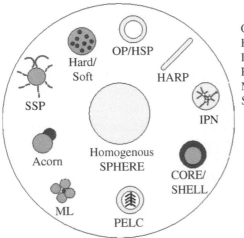

OP/HSP –Opaque Polymer
HARP –High Aspect Ratio Polymer
IPN –Interpenetrating Network
PELC –Polymer Encapsulated Latex
ML –Multilobe^{TM}
SSP –Soluble Shell Polymer

Figure 8.14   Morphologies from Rohm and Haas Company.

Dr John M. Friel
Rohm & Haas
Research Laboratories
Philadelphia
USA

SAMPLE PROBLEMS

(1) Calculate the Debye lengths for 0.1 mM, 10 mM and 100 mM aqueous solutions of NaCl and $MgSO_4$, assuming that the salts are completely ionized.

(2) Calculate and graph the concentration profiles for Na^+ and Cl^- ions next to a planar charged surface with a potential of −85 mV immersed in a 10 mM NaCl solution.

(3) Show graphically how the surface charge density varies with the surface potential for a planar surface in different Debye length solutions.

(4) Use the superposition principle to calculate the electrostatic swelling pressure generated between parallel clay platelets with surface potentials of −110 mV at a separation of 35 nm in a 1.5 mM aqueous solution of NaCl at 25 °C.

(5) Estimate the thickness of a water film of 0.1 mM NaCl solution on a glass plate 1 cm above a water reservoir. Assume that the water completely wets the glass, that the water/glass interface has an electrostatic potential of −60 mV and that only gravitational and electrical double-layer forces need be considered. Also assume that there is no surface charge at the water/air interface.

(6) The surface of a colloid dispersed in aqueous salt solution was found to have an equilibrium surface electrostatic potential of +80 mV, due to the specific adsorption of Na^+ ions. What is the (chemical) free energy of adsorption of Na^+ ions to this surface?

(7) Simplify equation (8.13) to give a simple physical interpretation of the Debye length.

(8) Estimate the concentration of Na^+ ions at the centre layer of an aqueous soap film drawn from a mM NaCl solution if the electrostatic potential at each surface is −20 mV and the film is 10 nm thick.

(9) The electrostatic potential at a distance of 5 nm away from a charged flat surface was found to be −10 mV in an aqueous 0.1 mM NaCl solution. Estimate the electrostatic potential at the surface. What are the concentrations of each ion at this distance away from the surface? Estimate the osmotic pressure at this plane.

Experiment

Zeta Potential Measurements at the Silica/Water Interface

Introduction

The stability of most colloidal solutions depends critically on the magnitude of the electrostatic potential (ψ_o) at the surface of the colloidal particle. One of the most important tasks in colloid science is therefore to obtain an estimate of ψ_o under a wide range of electrolyte conditions. In practice, the most convenient method of obtaining ψ_o uses the fact that a charged particle will move at some constant, limiting velocity under the influence of an applied electric field. Even quite small particles (i.e. < 1 μm) can be observed using a dark field illumination microscope and their (average) velocity directly measured. The technique is called microelectrophoresis.

At low electric fields [0(1V/cm)] the speed (U) of the particles is directly proportional to the applied field (E) and hence we can define a parameter called the electromobility (μ) of the particles, given by U/E. Using the Poisson-Boltzmann theory of the diffuse electrical layer next to a charged surface, a simple relationship between μ and ψ_o can be derived. However, because of doubts about the validity of the theory, we introduce another surface potential called the zeta potential (ζ) to represent the surface potential obtained from the electromobility measurements. This potential corresponds to the electrostatic potential at the plane of shear in the liquid, which is assumed to be close to the particle's surface. (Note that in general it usually reasonable to assume that $\psi_o = \zeta$).

This potential is given by the Smoluchowski equation:

$$\zeta = \frac{\mu\eta}{\varepsilon_o D} \tag{1}$$

where η is the viscosity of the solution, ε_o the permittivity of free space and D the dielectric constant. It is important to note that because of the method of derivation of this equation, it is only valid for colloidal particles which are large compared with the Debye length (κ^{-1}) of the electrolyte solution (i.e. $\kappa a >> 1$, where a is the radius of the colloidal particles). In general, therefore, this equation will be valid for particles of 1 μm or larger radii.

The surface of a silica (or glass) particle contains a high density of silanol groups (about 1 per 25 $Å^2$) which dissociate to some extent in water to give a negatively charged surface, thus:

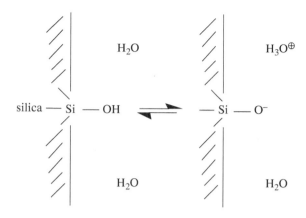

Figure 8.15 Ionisation of the surface of silica in water.

The magnitude of the electrostatic potential at the silica surface is, as expected from the law of mass action, pH dependent. The variation in surface (or zeta) potential with pH must therefore be dependent on the dissociation constant of the surface silanol (Si-OH) group.

In this experiment a zeta meter is used to determine the variation in the zeta potential of silica at constant pH (5.7) over a range of concentrations of a cationic surfactant CTAB, which should adsorb on the ionized silanol groups on the silica particle surface.

Experimental procedures

This experiment is based on the use of a classic Rank Bros (Cambridge, UK) Mark 2 microelectrophoresis apparatus, which is based on the manual microscopic observation of the speed of colloids, detected by dark field illumination, under the action of an applied electric field. Either a rectangular or cylindrical quartz cell can be use. For a comprehensive reference text, see R. J. Hunter, *The Zeta Potential* (London: Academic Press, 1981). Before using the zeta meter apparatus, make up colloidal solutions of silica using about 0.01 gm solid per 100 ml of a range of CTAB solutions from 10^{-6} M to 10^{-2} M, in 10^{-2} M KBr (to keep the Debye length constant). Always shake these solutions thoroughly and maintain at a temperature of 30 °C before transferring to the apparatus cell, which is also set at 30 °C. This is because CTAB solutions have a Krafft temperature or solubility point at the critical micelle concentration (cmc) at about 25 °C.

An illustration of the illumination used in this type of apparatus to observe the motion of colloid particles is given in the following figure:

Dark-field illumination (zeta meter)

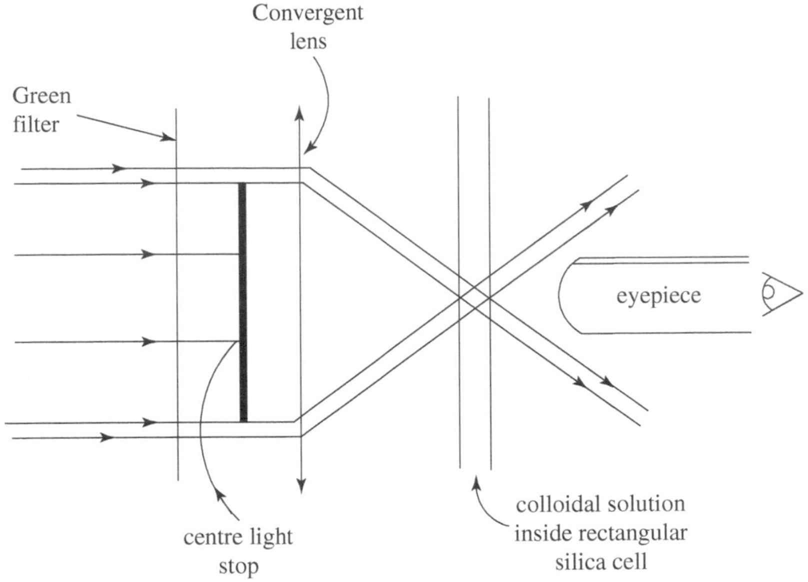

Figure 8.16 Diagram of the dark-field illumination system used to visualise colloidal particles dispersed in water.

Basically, the illumination system and microscope allow you to observe the motion of the silica particles, which are seen as bright star-like objects on a green bac-ground. When an electric field is applied the average time taken for the particles to travel a distance of one square on the eyepiece graticule can be easily measured. One particle is measured each time the field is applied for a short time (i.e. 10–20 s) and the polarity is then reversed and the speed in the opposite direction measured. The polarity must be reversed *each time* and the field never left on for longer than about 30 seconds, so that the possibility of polarization effects is reduced. Usually, between 10 and 20 particles in the field of view are measured and the average value obtained. The applied voltage should be varied to make the particles move over one square in about 10 seconds, but this voltage must never be increased above about 30 V. Be careful to measure only particles clearly in the plane of focus, since the microscope will have been set to measure at the fluid stationary plane within the rectangular cell. Particles not in focus will travel at a speed which will include a fluid flow component and will lead to experimental errors if included in the data.

A photograph of a typical Rank Bros instrument is given below:

Quartz cell   Lamp source

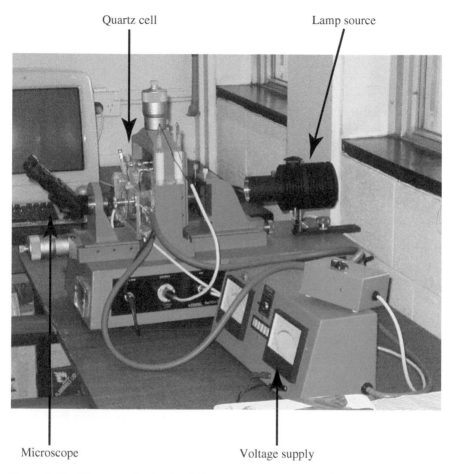

Microscope   Voltage supply

Figure 8.17   Photograph of a Rank Bros MK 2 microelectrophoresis instrument.

Once the apparatus is set-up, all you have to do is change the colloidal solutions in the rectangular cell and measure their mobilities. Be careful to check the direction the particles move, which denotes their sign of charge. This sign will change at some CTAB concentration. (As a guideline, in water the silica particles are negatively charged).

In order to change solutions, the electrodes are first carefully removed (very little effort should be required to do this – be careful; these cells are expensive) and are then stored in distilled water. The old solution is aspirated out and the cell rinsed and aspirated with distilled water. Finally, the new solution is poured into the cell and the electrodes carefully replaced so that no air bubbles are trapped near the electrodes. The new solution should be left to equilibrate (at 30 °C) for about 15 minutes before measurement. A typical rectangular quartz cell is shown below:

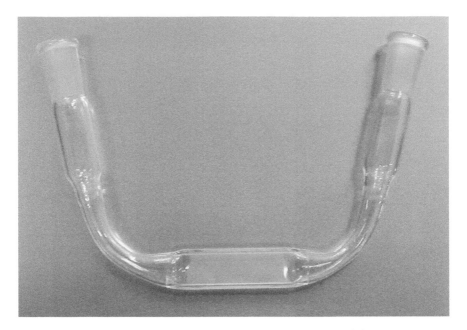

Figure 8.18 Rectangular quartz cell used to measure electromobility.

The speed of the colloid particles is measured at a stationary fluid layer within the rectangular cell. These cells are expensive and must be handled carefully. At the end of the experiment, the electrodes should be removed and placed in distilled water, and the cell should be rinsed and filled with distilled water.

The speed of the particles can be easily calculated using the fact that each square of the graticule corresponds to a known distance, typically 60 μm. The electric field applied is also simply calculated by dividing the applied voltage directly measured on the instrument by the distance between the electrodes. The effective inter-electrode distance is obtained by measuring the conductance of a standard electrolyte solution, say 0.01 M KCl, in the cell. The distance is calculated from the known conductivity of the solution. The average electromobility of the colloids is thus obtained and hence, using equation 1, the average zeta potential.

Plot your results on a graph of zeta potential in mV as a function of CTAB concentration.

FOR CONSIDERATION/TYPICAL QUESTIONS

(1) Your basic data is in the form of time (in seconds). In determining the average speed of the particles, should you first average the time intervals and then invert to calculate the average speed, or should you invert the times (to give the speed) and then find the average speed?

(2) At what concentration of CTAB was the silica surface uncharged?

(3) Draw a schematic diagram of the type of CTAB adsorption you would expect at each concentration.

(4) Would you expect CTAB adsorption to increase further with concentrations above 10^{-3} M?

9

Van Der Waals Forces and Colloid Stability

Historical development of van der Waals forces. The Lennard-Jones potential. Intermolecular forces. Van der Waals forces between surfaces and colloids. The Hamaker constant. The DLVO theory of colloidal stability.

HISTORICAL DEVELOPMENT OF VAN DER WAALS FORCES AND THE LENNARD-JONES POTENTIAL

In 1873 van der Waals pointed out that real gases do not obey the ideal gas equation $PV = RT$ and suggested that two 'correction' terms should be included to give a more accurate representation, of the form $(P + a/v^2)$ $(V - b) = RT$. The term a/v^2 corrects for the fact that there will be an attractive force between all gas molecules (both polar and non-polar) and hence the observed pressure must be increased to that of an ideal, non-interacting gas. The second term (b) corrects for the fact that the molecules are finite in size and act like hard spheres on collision; the actual free volume must then be less than the total measured volume of the gas. These correction terms are clearly to do with the interaction energy between molecules in the gas phase.

In 1903 Mie proposed a general equation to account for the interaction energy (V) between molecules:

$$V = -\frac{C}{d^m} + \frac{B}{d^n}$$

(9.1)

Applied Colloid and Surface Chemistry, Second Edition. Richard M. Pashley and Marilyn E. Karaman.
© 2021 John Wiley & Sons Ltd. Published 2021 by John Wiley & Sons Ltd.
Companion website: www.wiley.com/go/pashley/appliedcolloid2e

of which the most usual and mathematically convenient form is the Lennard-Jones 6-12 potential:

$$V_{LJ} = -\frac{C}{d^6} + \frac{B}{d^{12}}$$

(9.2)

where the first term represents the attraction and the second the repulsion between two molecules separated by distance d. This equation quite successfully describes the interaction between non-polar molecules, where the attraction is due to so-called dispersion forces, and the very short-range second term is the Born repulsion, caused by the overlap of molecular orbitals.

From our observation of real gases, it is clear that attractive dispersion forces exist between all neutral, non-polar molecules. These forces are also referred to as London forces after the explanation given by him in about 1930. At any given instant, a neutral molecule will have a dipole moment because of fluctuations in the electron distribution in the molecule. This dipole will create an electric field which will polarize a nearby neutral molecule, inducing a correlated dipole moment. The interaction between these dipoles leads to an attractive energy of the form $V = -C/d^6$. The time-averaged dipole moment of each molecule is, of course, zero, but the time-averaged interaction energy is finite, because of this correlation between interacting temporary diploes. It is mainly this force which holds molecular solids and liquids, such as hydrocarbons and liquefied gases, together. The L-J interaction potential V between molecules of a liquid (or solid) separated by distance d is illustrated below:

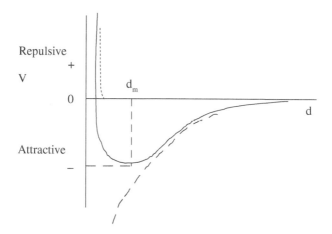

Figure 9.1   Interaction energy between two molecules.

In this case the molecules in the liquid would have an equilibrium spacing of d_m. Further, if we knew the detailed 'structure' of the liquid, that is, the radial distribution function g(r), we could calculate its internal energy <U> and hence relate molecular interactions directly to the system's thermodynamic parameters. This calculation can be performed using the classical Hamiltonian function of the sum of the internal (interaction) energy and the molecular kinetic energy, $3NkT/2$, shown below:

$$\langle U \rangle = \frac{3}{2}NkT + \frac{1}{2}N\rho\int_0^\infty 4\pi r^2 g(r)V(r)dr$$

(9.3)

and if we know how g(r) varies with temperature, then we can calculate <U> as a function of temperature, from which other thermodynamic parameters can be generated.

It is interesting to note that molecular interaction forces vary with the inverse distance to a power greater than 3. This means that these forces are short-ranged within a material, unlike gravitational forces, and only molecules within 10 or so diameters effectively contribute to the cohesive or surface energy of a material. This is quite unlike gravity, which has a slower distance dependence and so has to be summed over the entire body or region of space. Interestingly, this difference was not appreciated by Einstein in some of his early work.

Because the dispersion force acts between neutral molecules, it is ubiquitous (compare the gravitational force); however, between polar molecules there are also other forces. Thus, there may be permanent dipole-dipole and dipole-induced dipole interactions, and of course, between ionic species there is the Coulomb interaction. The total force between polar and non-polar (but not ionic) molecules is called the van der Waals force. Each component can be described by an equation of the form $V = C/d^n$, where for the dipole-dipole case $n = 6$ and C is a function of the dipole moments. Clearly, it is easy to give a reasonable distance dependence to an interaction; however, the real difficulty arises in determining the value of C.

Common types of interactions between atoms, ions and molecules in vacuum are given in the table below. In the table, V(r) is the interaction free energy (in J); Q is the electric charge (in C); u the electric dipole moment (C m); α the electric polarisibility ($C^2m^2J^{-1}$) and r is the distance between interacting atoms or molecules (m). v, is the electronic absorption (ionization) frequency (s^{-1}). The corresponding interaction force is, in each case, obtained by differentiating the energy V(r) with respect to distance r.

Table 9.1

Type of Interaction	Interaction energy $V(r)$
Covalent	Complicated, short range
Charge-charge	$\dfrac{q_1 q_2}{4\pi\varepsilon_0 r}$ Coulomb's law
Charge (q) – dipole (u), freely rotating dipole:	$\dfrac{-q^2 u^2}{6\left(4\pi\varepsilon_0\right)^2 kTr^4}$
Dipole-dipole, both freely rotating:	$\dfrac{-u_1^2 u_2^2}{3\left(4\pi\varepsilon_0\right)^2 kTr^6}$ (Keesom energy)
Charge (q) – non-polar (α):	$\dfrac{-q^2 \alpha}{2\left(4\pi\varepsilon_0\right)^2 r^4}$
Rotating dipole (u) – non-polar (α):	$\dfrac{-u^2 \alpha}{\left(4\pi\varepsilon_0\right)^2 r^6}$ (Debye energy)
Two non-polar molecules:	$\dfrac{-3h\nu\alpha^2}{4\left(4\pi\varepsilon_0\right)^2 r^6}$ (London dispersion energy)
Hydrogen bond:	Complicated short range energy roughly proportional to $-1/r^2$

A more sophisticated model for water molecule interactions is given by the four-point charge model. The interaction potential has the form:

$$V\left(X_1, X_2\right) = V_{LJ}\left(R_{12}\right) + S\left(R_{12}\right) V_{HB}\left(X_1, X_2\right) \tag{9.4}$$

where X_1 is the generalized coordinate of molecule 1 specifying the position and orientation of that particle. $V_{LJ}(R_{12})$ is a Lennard-Jones 6:12 potential with parameters for neon, which is iso-electronic with water, and V_{HB} is a slight variant of the Bjerrum four-point charge model (below). The function $S(R_{12})$ is a switching function to give small weight to configurations in which the two point charges on neighbouring molecules overlap.

The modified four-point-charge model of Bjerrum has been used for molecular dynamics simulations of liquid water to calculate its thermodynamic and structural properties. It turns out that this very important liquid is extremely difficult to model theoretically. This sort of model can be used to compute the interaction of between 100 to 1000 molecules for a simulation time of about a ps or 10^{-12} seconds! That there are about 50 different

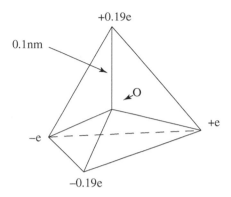

+0.19e

0.1nm

O

+e

−e

−0.19e

Figure 9.2   Bjerrrum four-point-charge model for water.

water models indicates that most are not completely successful in modelling all the liquid properties of water.

DISPERSION FORCES

It is instructive to follow the derivation of the London dispersion interaction for the simplest case of two interacting hydrogen atoms, using the Bohr model where the electron is regarded as travelling in well-defined orbits about the nucleus. The orbit of smallest radius, a_0, is the ground state, and Bohr calculated that

$$a_0 = \frac{q^2}{8\pi\varepsilon_0 h\upsilon}$$

(9.5)

where υ is the characteristic frequency (in Hz) associated with the electron's motion around the nucleus. (Note that the value of a_0 corresponds to the maximum value in the electron density distribution $|\psi|^2$ in the electronic ground state of hydrogen as calculated from quantum mechanics.) The energy $h\upsilon$ is the first ionisation potential of the H atom. Consider the H atom as having a spontaneously formed dipole moment

$$P_1 = a_0 q$$

(9.6)

The electric field E due to this instantaneous dipole at distance R will be:

$$E \cong \frac{P_1}{4\pi\varepsilon_0 R^3} \cong \frac{a_0 q}{4\pi\varepsilon_0 R^3}$$

(9.7)

(which can be calculated simply from Coulomb's law). If a neutral molecule is at R, it will be polarized, producing a dipole moment depending on its polarizability α, thus

$$P_2 = \alpha E \cong \frac{\alpha a_0 q}{4\pi\varepsilon_0 R^3}$$

(9.8)

α measures the ease with which the electron distribution can be displaced and is proportional to the volume of the atom (H in this case):

$$\alpha \cong 4\pi\varepsilon_0 a_0^3$$

(9.9)

The potential energy of interaction of the dipoles P_1, and P_2 is then:

$$V(R) = -\frac{P_1 P_2}{4\pi\varepsilon_0 R^3} \left(\text{for fixed and aligned diploes}\right)$$

$$= -\frac{\alpha a_0^2 q^2}{\left(4\pi\varepsilon_0\right)^2 R^6}$$

$$\therefore V(R) = -\frac{2h\upsilon\alpha^2}{\left(4\pi\varepsilon_0\right)^2 R^6}$$

(9.10)

Thus, the dispersion force interaction depends very much on the polarizability or response of the molecule to an electric field.

RETARDED FORCES

Before we move on to consider the interaction between macroscopic bodies, let us look briefly at the phenomenon of retardation. The electric field emitted by an instantaneously polarised neutral molecule takes a finite time to travel to another neighbouring molecule. If the molecules are not too far apart, the field produced by the induced dipole will reach the first molecule before it has time to disappear or perhaps form a dipole in the opposite direction. The latter effect does, however, occur at larger separations (>5 nm) and effectively strengthens the rate of decay with distance, producing a dependence of $1/R^7$ instead of $1/R^6$. At closer separations, where the van der Waals forces are strong, the interaction is non-retarded, and we will assume this is the case from here onwards.

VAN DER WAALS FORCES BETWEEN MACROSCOPIC BODIES

In colloid and surface science we are interested in calculating the van der Waals interaction between macroscopic bodies, such as spherical particles and planar surfaces. If the dispersion interaction, for example, were additive, we could sum the total interaction between every molecule in a body with that in another. Thus, if the separation distance between any two molecules i and j in a system is

$$R_{ij} = \left| \vec{R}_j - \vec{R}_i \right|$$

(9.11)

then a sum of the interaction energy between all molecules gives:

$$V^{1,2,\ldots N} = 0.5 \sum_{i=1}^{N} \sum_{j=1(\neq i)}^{N} V^{i,j}\left(R_{ij}\right)$$

(9.12)

where $V^{ij}(R_{ij})$ is the interaction energy between molecules i and j in the *absence* of all other molecules. However, this approach is only an approximation because the interaction is dependent on the presence of other molecules, and the correct treatment has to include many-body effects. A complete but rather complicated theory which includes these effects is called the Lifshitz theory, which was derived via quantum field theoretic techniques. Although the complete equation is complicated, we can represent it in a simple and approximate form when the interaction is non-retarded (at separations less than about 5 nm). Thus, for the case of two planar macroscopic bodies:

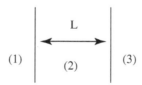

Figure 9.3   Diagram of two planar surfaces separated by distance L.

The van der Waals interaction energy per unit area is given by:

$$V_{123}^{F}(L) = -\frac{A_{123}}{12\pi L^2}$$

(9.13)

where A_{123} is the so-called Hamaker constant, which is positive when the interaction is attractive. Similarly, between identical spherical particles:

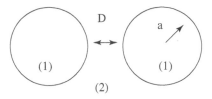

Figure 9.4   Diagram of two colloidal spheres separated by distance D.

$$V_{121}^S(D) = -\frac{A_{121}a}{12D} \tag{9.14}$$

These equations are simple and become powerful if we know the value of A for particular materials. Tables of A values which have been calculated using the Lifshitz theory are available, and some examples are given below:

Table 9.2

Components	Hamaker constant
	$A\ (10^{-20}\ \mathrm{J})$
water/vacuum/water	3.7
polystyrene/vacuum/polystyrene	6.1–7.9
silver/vacuum/silver	40
quartz/vacuum/quartz	8–10
hydrocarbon/vacuum/hydrocarbon	6
polystyrene/water/polystyrene	1.3
quartz/water/quartz	1.0
dodecane/water/dodecane	0.5
Teflon/water/Teflon	0.3

THEORY OF THE HAMAKER CONSTANT

Examination of this table reveals some interesting features, such as the effect of the medium in between two macroscopic bodies, which clearly has a marked effect on reducing the van der Waals attraction. This effect can be understood if we examine the difference between bringing a second body towards another in a vacuum and when immersed in another dielectric medium. In the first case, the original body has no molecules to interact with outside itself and hence lower its free energy, whereas in the second the body already interacts with molecules in the surrounding medium, which are merely replaced by those in the approaching second body. The magnitude of the Hamaker constant in the second case will clearly depend on the interaction between all the components. This is also the reason why any two dielectrics will be attracted by van der Waals forces in a vacuum but not necessarily when immersed in some other medium.

Although it is reasonable to simply use the values calculated by theoreticians (and in a few cases measured by experimentalists) for the Hamaker constant, it is important to understand something about how it is calculated. The Hamaker constant is, in fact, a complicated function of the frequency-dependent dielectric properties of all the media involved. The way in which the varying electric fields generated by one body interact with another determines the van der Waals interaction. In order to understand this, let us look at the effect of placing a dielectric material (i.e. $\varepsilon > \varepsilon_0$) between two charged metal (conducting) plates initially in a vacuum. This situation is illustrated below:

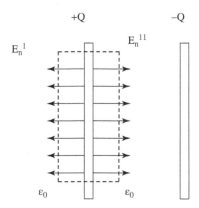

Figure 9.5 Electric field around two charged plates of a capacitor.

One of the fundamental laws of electrostatics is that due to Gauss:

$$\int \vec{E} \cdot \hat{n}\, da = \frac{Q}{\varepsilon_0}$$

$$(9.15)$$

which may be derived from Maxwell's equations (see earlier) and says that the total electric flux through (normal to) a surface is directly related to the total charge Q inside the surface. Let us use this general equation to study one half of the capacitor system illustrated, where the total area is shown as a dashed line (above).

The electric flux normal to the surface is given by:

$$\left(E'_n + E''_n \right) = \frac{Q}{A\varepsilon_0}$$

$$(9.16)$$

where A is the area of one side of the capacitor plate and Q is the total charge on the plate. However, since the electric fields are symmetrical around a flat plate $E'_n = E''_n$ and hence $E = Q/2A\varepsilon_0$. If we now also include the field due to the other negative plate, this becomes:

$$E = Q / A\varepsilon_0$$

$$(9.17)$$

Since the capacitance, C, is defined as:

$$C = Q / V$$

$$(9.18)$$

where V is the potential difference between the plates, and this is the work done in moving a unit positive charge from one plate to the other, that is:

$$Ed = V$$

$$(9.19)$$

(since the field E is the force acting on a unit positive charge inside the capacitor plates) we can obtain the simple result that:

$$C_0 = \frac{\varepsilon_0 A}{d}$$

$$(9.20)$$

where the capacitance depends only on the area and the separation of the plates and the permittivity of the medium between the plates. What happens if we now fill this free space with a dielectric material? Since for any dielectric $\varepsilon_D > \varepsilon_0$, we can see immediately that the capacitance will increase via the formula:

$$\frac{C_D}{C_0} = \frac{\varepsilon_D}{\varepsilon_0}$$

$$(9.21)$$

This is a fundamental property of dielectrics and means that if the charge on the plates is fixed, the potential difference between the plates and hence the electric field inside (in the dielectric) must have fallen compared with that in free space. How this happens is explained in the following schematic figure:

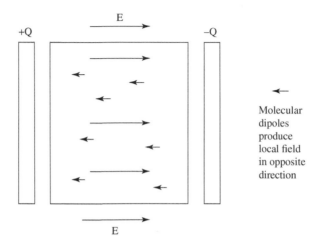

Figure 9.6 Effect of a dielectric material on the electric field within a capacitor.

In the electric field between the plates, the dielectric material polarises, for example, by a change in electron distribution in the molecules or, if polar, by reorientation of the dipoles (especially for a liquid such as water). The induced dipolar field then must act in the opposite direction and reduces the total electric field and the potential difference. Thus, the dielectric material between the plates of the capacitor increases its capacitance. The dielectric constant of the material can be measured this way.

Now let us examine what would happen to the response of the dielectric if we put an alternating voltage on the capacitor of frequency ω. If ω is low (a few Hz), we would expect the material to respond in a similar manner to the fixed voltage case, that is ε_D (static) $= \varepsilon(\omega) = \varepsilon(0)$. (It should be noted that ε_0, the permittivity of free space, is not frequency dependent and that $\varepsilon(0)/\varepsilon_0 = D$, the static dielectric constant of the medium.) However, if we were to increase ω to above microwave frequencies, the rotational dipole response of the medium would disappear, and hence $\varepsilon(\omega)$ must fall. Similarly, as we increase ω to above IR frequencies, the vibrational response to the field will be lost, and $\varepsilon(\omega)$ will again fall. Once we are above far UV frequencies, all dielectrics behave much like a plasma and eventually, at very high values, $\varepsilon(\omega)|_{\omega \approx \infty} = 1$.

What is actually happening at the specific frequencies ω_i, where there is a sudden reduction in the response of the dielectric? We can, in fact, treat the interaction of varying electric fields with dielectrics as though the latter were made up of electron or dipole oscillators, such that when the resonant frequency is reached, electric energy is absorbed and, usually, dissipated as heat. The specific ω_i values must then correspond to the absorption peaks of the material. In order to represent this behaviour, we allow the frequency dependent dielectric constant to have an imaginary component, thus:

$$\varepsilon(\omega) = \varepsilon'(\omega) + i\varepsilon''(\omega) \qquad (9.22)$$

where $\varepsilon''(\omega)$ is directly proportional to the absorption coefficient of the dielectric. At frequencies where there is no absorption it follows that:

$$\varepsilon(\omega) = \varepsilon'(\omega) = n(\omega)^2 \qquad (9.23)$$

where $n(\omega)$ is the refractive index (at that frequency). From what we have said about dielectrics, we would expect something like the following for the behaviour of $\varepsilon'(\omega)$ and $\varepsilon''(\omega)$, with frequency of the electric field:

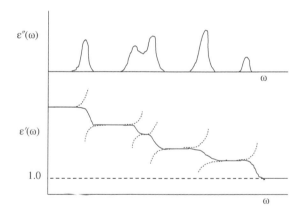

Figure 9.7 Typical responses for the real and imaginary components of the dielectric constant to frequency.

If the absorption data, that is $\varepsilon''(\omega)$, over a very wide frequency range (in practice from microwave to far UV) has been measured, we can use the Lifshitz equation to calculate the Hamaker constant. Unfortunately, this data is known in detail for only a few materials (such as water and polystyrene), and so oscillator models for the main absorption peaks (say in the IR and UV) are often constructed and used to calculate $\varepsilon''(\omega)$.

We can, however, make some semi-quantitative comments about the type of van der Waals forces we expect from the main absorption peaks and the refractive index of transparent dielectrics. For example, if two dielectric bodies which interact through vacuum have very similar absorption spectra, the van der Waals attraction will be strong. Also, if the intervening medium has a spectrum similar to the interacting bodies, the attraction will be weak (and can even be repulsive).

Even if we know nothing about the absorption properties, we can still deduce something about the magnitude of van der Waals forces from the refractive index at visible frequencies where it is known for transparent (i.e. non-absorbing) materials. This is because a high refractive index must mean that there is still a substantial dielectric response at these (visible) frequencies and therefore there must be further absorption at higher frequencies in the UV range (since $\varepsilon(\omega) = n(\omega)^2$ where the material is transparent). Absorption in the UV range is very important for short-range van der Waals interactions ($d < 5$ nm), where these frequencies are not retarded, hence the refractive index is related to the Hamaker constant. This is illustrated by the following table, which lists refractive indices of materials and the corresponding Hamaker constants for interactions across a vacuum:

Table 9.3

Material	n_D (λ=589.3 nm)	$A_{||}$ (10^{-20} J)
water	1.33	3.7
hexadecane	1.43	5.2
CaF_2	1.43	7.2
fused silica (SiO_2)	1.46	6.6
crystalline quartz (SiO_2)	1.54	8.8
mica (aluminosilicate)	1.55	10
calcite ($CaCO_3$)	1.5–1.7	10.1
sapphire (Al_2O_3)	1.77	16

As an aside, it should be noted that even when electromagnetic radiation (e.g. light) passes through a transparent dielectric medium, the varying electric fields polarise the molecules, and these re-emit radiation of the same wavelength but with a phase shift. It is the combination of this re-emitted field with the original which gives a transmitted beam an appearance of travelling at a slower speed, C_D, compared with its speed in a vacuum, C. The refractive index can be shown from this approach to be numerically equal to the ratio

C/C_D. The electric field 'in between' the molecules of the dielectric still travels with the speed of light in a vacuum. Absorption occurs when the interacting field is not re-emitted, and this occurs at the resonant frequencies of the material, where the dissipated electric field energy usually appears as heat.

USE OF HAMAKER CONSTANTS

Once we have established reasonable values for the Hamaker constants, we should be able to calculate, for example, adhesion and surface energies, as well as the interaction between macroscopic bodies and colloidal particles. Clearly, this is possible if the only forces involved are van der Waals forces. That this is the case for non-polar liquids such as hydrocarbons can be illustrated by calculating the surface energy of these liquids, which can be directly measured. When we separate unit area of a liquid in air, we must do work W_C (per unit area) to create new surface, thus:

$$V_{11} = -W_c = -\frac{A_{11}}{12\pi L^2} \tag{9.24}$$

If we assume a reasonable separation distance in the liquid, of say $L = 0.15$ nm, then we calculate a liquid surface tension value, $\gamma_{l\,(calc)}$, of 30 mJm^{-2}, whereas the measured value is 28 mJm^{-2}, which is reasonably close. However, if we carry out the same calculation for water using the same spacing value, then $\gamma_{l\,(calc)} = 22$ mJm^{-2}, which is much lower than the measured value of about 72 mJm^{-2}. The main reason for this large discrepancy is that the surface energy of water is high because of the short-range structural hydrogen-bonding between water molecules. At the water/air interface these molecules are orientated quite differently to those in bulk. Of course we have included dipole-dipole interactions in our calculation of the Hamaker constant (by using the microwave and zero frequency contributions), but these are the bulk properties of liquid water, which do not represent structural and orientational changes in the water dipoles at the surface of the liquid. Application of the theory to calculate surface tension values clearly works best for simple liquids.

THE DLVO THEORY OF COLLOID STABILITY

Although for some interfaces it can be difficult to calculate the surface energy this way, the Hamaker constant does afford us a powerful tool for the calculation of the attractive forces acting between colloids.

Typically, these forces are opposed by the charge on the particle surface and the generation of a repulsive stabilising force, called the electrical double-layer repulsion, discussed in Chapter 8. Combination of the simple equation obtained in the present Chapter (Equation 9.14) for the van der Waals attraction between spherical colloids, with that derived in Chapter 8 for the double-layer repulsion (Equation 8.12), leads to a simplified form of the DLVO theory for the stability of colloidal solutions. Thus, for the general case of two interacting spherical colloids illustrated below:

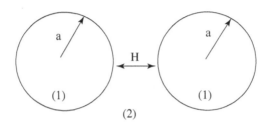

Figure 9.8   Two interacting identical colloidal particles.

the total interaction energy can be given approximately by the relation:

$$V_T \approx \left[2\pi a \varepsilon_o D \psi_o^2 \exp(-\kappa H) \right] - \frac{aA_{121}}{12H} \qquad (9.25)$$

Some typical interaction curves are given below:

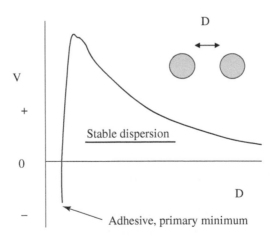

Figure 9.9   Some typical DLVO interaction curves.

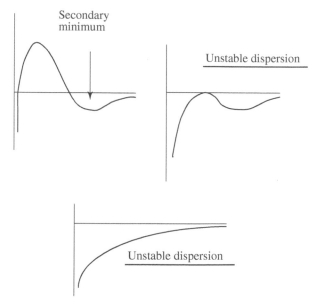

Figure 9.9 (*Continued*)

One of the first accurate experimental tests of the DLVO theory was published by one of us (RMP) in 1981 (in the *Journal of Colloid and Interface Science*, 80: 153). To obtain the required accuracy, the forces were measured not between colloids but between two molecularly smooth crystals of muscovite mica. The results obtained were in almost perfect agreement with theory:

The measured experimental points (large black dots in the figure, actually values of the interaction force F scaled by the radius of the surfaces R) were found to agree with the DLVO predictions (dashed line) at all separations. The long range of the forces was due to the low electrolyte concentration in distilled water, giving an effective Debye length of 145 nm. At small separations the inset shows that the surfaces were pulled into adhesive contact by van der Waals forces, as predicted by the theory. These results strongly supported the validity and accuracy of the DLVO theory applied to surface interactions in dilute solutions. Later work suggests that the theory is less accurate in more concentrated solutions, where other more ion-specific affects occur, such as effects due to ion hydration, and this has led to much debate. Unfortunately, many real-world interactions occur in relatively concentrated electrolyte solutions, such as in seawater or in the human body. Under these conditions, specific ion-solvent effects also have to be considered, rather than a simple point ion, average-field continuum model, which is the basis of the DLVO theory. Surface interactions in concentrated solutions continues to be an active area of research.

In 1991, one of us (RMP) also helped to develop a new experimental procedure, called the colloid probe technique, which is now widely used to

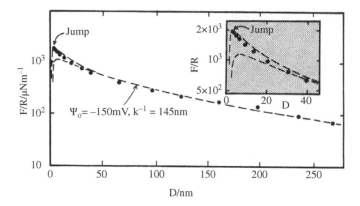

Figure 9.10 Measured DLVO forces between two molecularly smooth mica surfaces in water.

measure the interaction forces between colloidal surfaces (see W. A. Ducker, T. J. Senden and R. M. Pashley, *Nature*, 353: 239–241) using an atomic force microscope (or AFM). Following is a photograph of the original silica sphere colloid attached to an AFM cantilever. The interaction forces between this colloid and a flat silica plate were measured in a range of electrolyte solutions also to test the DLVO theory.

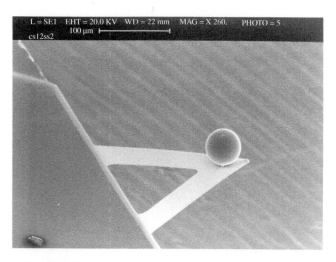

Figure 9.11 The first time an AFM was used to measure surface forces between a colloidal particle and a second surface in aqueous solution.

For colloidal solutions, as a general rule, a barrier of 15–25 kT is sufficient to give colloid stability, where the Debye length is also relatively large, say, greater than 20 nm. This electrostatic barrier is sufficient to maintain metastability of the colloidal dispersion, but the system is usually not thermodynamically stable and even a low solubility of the dispersed phase will allow growth of large particles at the expense of small ones, as expected from the interfacial energy at the particle surface. This growth is called Ostwald ripening. The rate at which it occurs will depend on the solubility of the disperse phase in the dispersion medium. Note that even silica has a finite solubility in water.

If the electrostatic barrier is removed either by specific ion adsorption or by addition of electrolyte, the rate of coagulation (often followed by measuring changes in turbidity) can be described fairly well from simple diffusion-controlled kinetics and the assumption that all collisions lead to adhesion and particle growth. Overbeek (in H. R. Kruyt, *Colloid Science*, New York: Elsevier, 1952) has derived a simple equation to relate the rate of coagulation to the magnitude of the repulsive barrier. The equation is written in terms of the stability ratio:

$$W = \frac{R_f}{R_s}$$

where R_f is the rate of fast coagulation, i.e. where there is no barrier and all collisions are successful, and R_s is the rate of slow coagulation against a barrier. The dependence of W on the barrier is given by the relation:

$$W = \frac{1}{2\kappa a} \exp\left(\frac{V_{max}}{kT}\right)$$

Thus, for a barrier of 20 kT and a κa value of 1, the stability ratio, W, has a value of 2.4×10^8, which corresponds to a slow rate of coagulation.

Stability from coagulation is an important property in many industrial processes and products. Often the electrostatic component is not sufficient, for example in high electrolyte solutions, and it is necessary to generate an additional repulsive barrier between the particles. One commonly used method is to adsorb a water-soluble polymer as a coating around the particles to prevent their close approach. The forces generated are complex and polymer specific but are classified as steric forces, to distinguish them clearly from the DLVO forces. This is the main reason many biological cells have coatings of natural biopolymers, to give stability in the high level of aqueous electrolyte in animals, of about 0.15 M NaCl.

An important industrial example of one of the common types of interaction obtained using this equation is the case of latex particles stabilised by repulsive electrostatic forces. The case shown below reflects the situation where rapid coagulation will be generated by even a small increase in electrolyte level by evaporation of water from the paint.

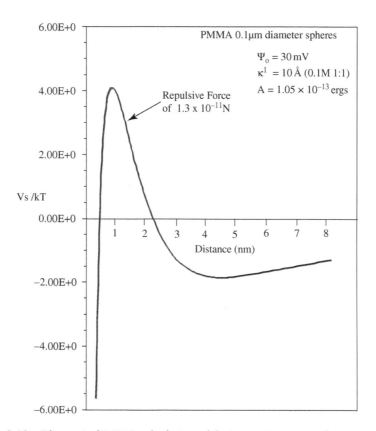

Figure 9.12 Theoretical DLVO calculation of the interaction energy between typical latex particles in water-based paints.

Whether stabilised by repulsive electrostatic forces or steric forces, the stable latex paint dispersion must rapidly become unstable when applied as a thin film coating. Typically, the initial latex dispersion contains 50% polymer particles with a diameter of about 0.1 microns, prepared by the emulsion polymerisation process described in Chapter 7. Once coated as a thin film, the evaporation of water forces the particles together into a close-packed array, then a compressed state and finally a continuous polymer film as illustrated below:

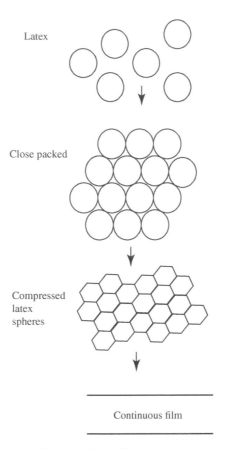

Latex

Close packed

Compressed
latex
spheres

Continuous film

Figure 9.13 Schematic diagram of the film formation process of latex paints.

An example of the surface of a drying latex film is shown in the following AFM image:

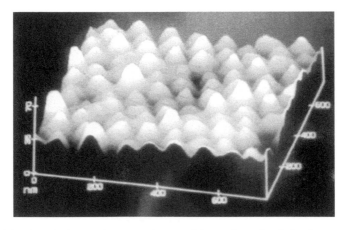

Figure 9.14 Atomic force microscope image of the surface of a drying latex paint film.

Both secondary and primary minimum coagulation are observed in practice and the rate of coagulation is dependent on the height of the barrier. In general, coagulation into a primary minimum is difficult to reverse, whereas coagulation into a secondary minimum is often easily reversed, for example, by diluting the electrolyte. DLVO theory tells us that colloidal solutions will, in general, be stable if the surface potential is high and the electrolyte concentration low, which is well supported by numerous experimental studies. The importance of the concentration and type of electrolyte in determining the stability of colloidal solutions is illustrated in the following experimental results:

Table 9.4   Critical Coagulation Concentrations (in mM).

As$_2$S$_3$(negatively charged)		Fe$_2$O$_3$(positively charged)	
Electrolyte	Coagulation concentration	Electrolyte	Coagulation concentration
NaCl	51	NaCl	9.25
KCl	49.5	KCl	9.0
KNO$_3$	50	(1/2)BaCl$_2$	9.6
(1/2)K$_2$SO$_4$	65.5	KNO$_3$	12
LiCl	58		
MgCl$_2$	0.72	K$_2$SO$_4$	0.205
MgSO$_4$	0.81	MgSO$_4$	0.22
CaCl$_2$	0.65	K$_2$Cr$_2$O$_7$	0.195
AlCl$_3$	0.093		
Al(NO$_3$)$_3$	0.095		
Ce(NO$_3$)$_3$	0.080		

One of the most effective methods for the coagulation and removal of colloids from solution arises from the use of heterocoagulation, where particles of opposite surface charge are mixed together. A very common and widely used industrial example is afforded by the processing of municipal drinking water supplies. The most common process uses a coagulant colloid in the form of an initially soluble salt, usually either ferric chloride or alum, which at moderate pH (~8) will hydrolyse to form positively charged flocs. Since almost all natural impurities in reservoir water are negatively charged, e.g. silica, clay and biological cells, the flocs attach to the contaminants, and the growing particles are settled and filtered out using high flow-rate sand bed filters.

Some notes on van der Waals forces

The degree of correlation in dielectric response between interacting materials leads to some useful generalisations about macroscopic van der Waals forces, thus:

(a) The vdw force is always attractive between any two materials in a vacuum. This is because there is no interaction between a dielectric material and a vacuum. However, in the Casimir effect, the spatial restriction of vacuum quantum fluctuations when two metal plates are placed in close proximity creates an attractive pressure on them in addition to the vdw force.

(b) The vdw force between identical materials is always attractive in any other medium. This is because the correlation will always be a maximum for identical materials – since different absorption spectra must always reduce the extent of correlation for different materials.

(c) The vdw force can be repulsive for two different materials interacting within a third medium. In this case one material must interact more strongly with the medium than with the second body.

Industrial Report

Surface chemistry in water treatment

In some instances the raw water reaching water treatment plants may contain pathogens such as the human infectious protozoa *Cryptosporidium parvum*. The environmental form of *C. parvum* is a spheroidal oocyst of 4–6 microns diameter. The oocyst is resistant to conventional chemical disinfectants that are commonly used in water treatment such as chlorine or chloramines. It is therefore essential that *Cryptosporidium* be removed during the coagulation and filtration processes stage in the water treatment train where chlorine and chloramines are relied on as the only disinfectants.

Water treatment by either direct or contact filtration has become common practice for raw water with low turbidity (<3 NTU) and low colour. Simple metal salts such as alum or ferric chloride are added to plant inlet water. Hydrolysis takes place with the formation of hydroxylated species, which adsorb reducing or neutralizing the charge on the colloidal particles in the raw water, promoting their collision and the formation of flocs that settle or can be filtered out.

In most water filtration plants in Sydney (Australia), which treat approximately 2×10^9 litres/day of raw water, the chemical regimes typically include (i) ferric chloride (up to 5 mg/L), (ii) low molecular weight cationic polyelectrolyte as secondary coagulants and (iii) high molecular weight non-ionic polyacrylamide-based flocculants as filter aids. Optimisation of the chemical regime in water filtration plants almost invariably delivers acceptable water quality in terms of turbidity (of about 0.15 NTU), colour and trace metals that meets standard water quality guidelines.

Research is continuing on the main factors that influence oocyst flocculation, since variations in chemical dosing, water chemistry or oocyst characteristics could potentially lead to oocyst breakthrough in water treatment plants. In recent years researchers have proposed that the interactions between oocysts and different coagulants may be quite different (Bustamante et al., 2001)[1].

Much of the understanding of the appropriate doses of coagulants to use has been developed from empirical success in optimising water treatment. It is therefore of primary importance to better understand the interaction of oocysts with common coagulants and coagulant aids normally used in water treatment to allow operators to make better informed choices in dosing during changing raw water conditions and to assist in trouble-shooting should problems arise.

Dr Heriberto Bustamante
Research Scientist
Sydney Water Corporation
Sydney, Australia

SAMPLE PROBLEMS

(1) Use schematic diagrams to describe the influence of electrolyte concentration, type of electrolyte, magnitude of surface electrostatic potential and strength of the Hamaker constant on the interaction energy between two colloidal-sized spherical particles in aqueous solution. What theory did you use to obtain your description? Briefly describe the main features of this theory.

(2) Two uncharged dielectric materials (1and 2) are dispersed as equal sized, spherical colloidal particles in a dielectric medium (3). If the refractive indices at visible frequencies follow the series $n_1 > n_3 > n_2$, determine the relative strengths (and sign, i.e. if repulsive or attractive) of the three possible interactions. Explain your reasoning.

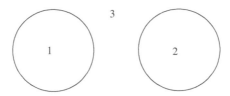

(3) Would you expect the van der Waals interaction between two spherical water droplets in air and two air bubbles in water to be the same if all the spheres are of identical size? Explain your answer.

(4) The total DLVO interaction energy (Vs) between two spherical colloids (each of radius a and separated by distance H) is given by the following approximate equation:

$$V_s = \left[2\pi a \varepsilon_o D \psi_o^2 \exp(-\kappa H)\right] - \frac{aA_{121}}{12H}$$

Calculate the critical value of the surface potential of the colloid which will just give the rapid coagulation case illustrated in the figure below. Assume that the aqueous solution contains 10 mM monovalent electrolyte at 25 °C. Also assume that the Hamaker constant for this case has a value of 5×10^{-20} J.

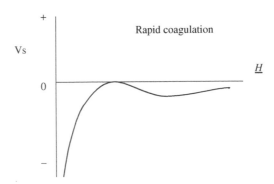

(5) Estimate the surface energy of liquid hexane using the fact that the van der Waals interaction for the liquid hexane/vacuum/liquid hexane (flat surface) case is given by:

$$V_{121} = -\frac{A_{121}}{12\pi L^2}$$

where the Hamaker constant A_{121} has a value of 6×10^{-20} J. Assume that the intermolecular spacing within liquid hydrocarbons is about 0.12 nm. How would you expect the Hamaker constant to change if the vacuum was replaced with (a) water and (b) another (immiscible) liquid hydrocarbon?

NOTE

1 Bustamante, H. A, Shanker, R, Pashley, R. M. and Karaman, M. E, 'Interaction Between *Cryptosporidium* Oocysts and Water Treatment Coagulants', *Water Research*, 35(13) 3179–3189, 2001.

10

Bubble Coalescence, Foams and Thin Surfactant Films

Thin liquid film stability. The effect of surfactants on film and foam stability. Surface elasticity. Froth flotation. The Langmuir trough and monolayer deposition. Laboratory project on the flotation of powdered silica.

THIN-LIQUID-FILM STABILITY AND THE EFFECTS OF SURFACTANTS

Foams are important in many everyday activities and are used in a diverse range of important industrial processes. In food science foams play a major role in both taste and appearance. Personal soaps contain compounds especially designed to stabilise foams, so that the soap can be both retained and easily transferred onto the skin during washing. Too much foam can also be a problem in industrial processes and in home washing machines. The major industrial process of froth flotation is based on the formation of foams for the collection and separation of valuable minerals in large quantities. Flotation is also used to remove earth during the cleaning and processing of vegetables.

When two air bubbles collide in water there is an overall thermodynamic advantage in fusion to form a larger single bubble. This follows directly from the reduced interfacial area and hence reduced interfacial energy:

Applied Colloid and Surface Chemistry, Second Edition. Richard M. Pashley and Marilyn E. Karaman.
© 2021 John Wiley & Sons Ltd. Published 2021 by John Wiley & Sons Ltd.
Companion website: www.wiley.com/go/pashley/appliedcolloid2e

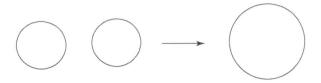

Figure 10.1 Illustration of the reduction in total surface area by the fusion of two air bubbles in water.

Since at constant gas volume the radius of the combined bubble must be 1.26 times the radius of the original bubble, it follows that there will be a reduction in interfacial area of just over 20%. This will be the case for all pure liquids and in fact the persistence of bubbles in water is often used as a simple check on its purity. Shaking a flask of water should produce bubbles that collapse within one to two seconds if the water is clean.

In order to understand the basis for the prevention of bubble coalescence and hence the formation of foams, let us examine the mechanical process involved in the initial stage of bubble coalescence. The relatively low Laplace pressure inside bubbles of reasonable size, say over 1 mm for air bubbles in water, means that the force required to drain the water between the approaching bubbles is sufficient to deform the bubbles as illustrated below:

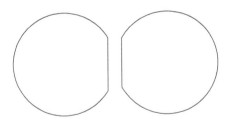

Figure 10.2 Deformation of rapidly colliding air bubbles in water.

The process which now occurs in the thin draining film is interesting and has been carefully studied. In water, it appears that the film ruptures, joining the two bubbles, when the film is still relatively thick, at about 100 nm thickness. However, van der Waals forces, which are attractive in this system (i.e. of air/water/air), are effectively insignificant at these film thicknesses. Most likely the rupturing occurs due to the correlation and fusion of surface waves at both surfaces of the film, as illustrated below:

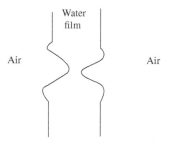

Figure 10.3   Surface correlated wave model to explain water film rupture.

It has been suggested that a long-range attractive force observed between hydrophobic surfaces in water, the so-called 'hydrophobic interaction', may also operate in thin water films to correlate the peaks in surface waves on the two facing surfaces and so induce fusion. Whatever the mechanism of film rupture, since the coalescence of bubbles is driven by surface energy changes, it is not surprising that adsorption of suitable 'surface active' materials will oppose fusion and create the phenomenon of foaming. Two main factors are necessary to create a stable foam:

(1) A repulsive force between the two film surfaces across the water film.
(2) A mechanism to induce surface elasticity in the film to protect against mechanical disturbances.

Both of these factors are achieved by the use of surfactants, which will adsorb at the two surfaces of the film, as illustrated below:

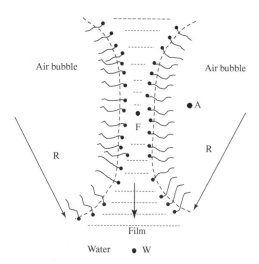

Figure 10.4   Surfactant adsorption at the surface of the bubbles stabilises the film and opposes film thinning, which then enables the creation of a stable foam structure.

The film (and hence the foam) can be stabilised by an electrostatic repulsive force acting between the ionised, adsorbed surfactants. It can also be stabilised by non-ionic polymeric type surfactants (e.g. ethylene oxide surfactants), where the large hydrophilic headgroups may have the effect of raising the viscosity of the film and hence reducing the rate of drainage. The film and foam are actually metastable because of the interfacial energy between the hydrocarbon monolayer and air. Also, as for colloidal solutions, van der Waals forces will tend to thin the film. However, foams can be further stabilized, even up to several months, by the addition of suitable polymers and partially hydrophobized nano particles.

THIN-FILM ELASTICITY

In the situation shown in the figure, we can calculate the pressure difference between points A and W using the Laplace equation: $\nabla P = \dfrac{2\gamma}{R}$. Across a flat interface there is no pressure difference, so the pressure in the film must be greater (by ∇P) than the pressure in the bulk liquid (at W). The pressure increase in the film is due entirely to the repulsive interaction between the two surfaces, which may be of electrostatic origin due to the ionisation of the surfactant head-groups. In addition to this increased pressure in the film, there is also a second factor which helps to give local stability to the foam. This is basically to do with the response of the film to mechanical shocks and thermal fluctuations, which of course must be withstood by a stable foam. Let us see what happens if a soap film is suddenly stretched during such a fluctuation, as illustrated below:

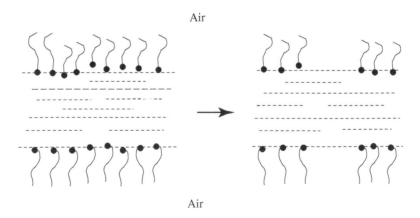

Figure 10.5   The effect of instantaneous stretching of a soap film.

A region must be instantaneously produced with exposed pure water interface before surfactant molecules have time to diffuse from bulk to the surface. Since the surface energy of the water region is much higher than of the surfactant region, there will be a strong surface restoring force to close up the exposed region. This effect gives rise to an apparent elasticity in the film which enhances mechanical stability.

A further elasticity factor arises directly from the properties of the surfactant film itself. Surface diffusion of adsorbed surfactant is much faster than diffusion from bulk (dilute) solution and from the data obtained in Langmuir-Blodgett studies (later in this chapter), increasing the area per head-group across the entire film surface, via stretching of the film, which will substantially increase the surface tension of the entire film when the surfactant layer is initially in the fully packed or compressed state. This non-local effect generates a substantial elasticity in the film given by $\frac{\partial \gamma}{\partial A}$ and protects the film from mechanical disturbances. An understanding of the mechanisms involved in producing stable foams leads directly to the methods used to destroy them. Surfactant foams can be destroyed simply by spraying with ethanol, which lowers the surface tension and allows the surface film to grow via its rapid diffusion from bulk to surface.

These various factors, when combined, lead to the everyday simple observation of soap foams, which have an almost infinite range of bubble sizes and combined shapes, as illustrated in the following photo.

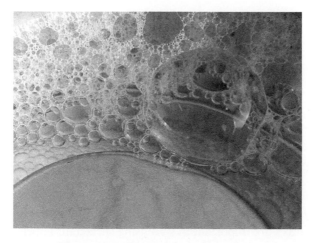

Figure 10.6 Typical foam formation.

REPULSIVE FORCES IN THIN LIQUID FILMS

The equilibrium thickness of a (meta-) stable soap film will depend on the strength and range of the repulsive forces in the film. Electrostatic forces are long range in water and hence give rise to thick (0.2 micron) films, which are highly coloured due to the interference of visible light reflected from the film. (It is for the same reason that oil films spilt on the road often show a variety of colours.) Films can also be stabilised by very short range (< 5 nm) steric and solvation forces. These are called Newton black films because visible light reflected from the front and back surface destructively interferes and so the films appear black. A beautiful range of colours can be observed by carefully allowing a large soap film, held on a rectangular metal frame, to slowly drain under gravity.

In this situation, the equilibrium thickness at any given height h is determined by the balance between the hydrostatic pressure in the liquid, ($h\rho g$) and the repulsive pressure in the film, that is: $\pi = h\rho g$. Cyril Isenberg gives many beautiful pictures of soap films of different geometry in his book *The Science of Soap Films and Soap Bubbles* (New York: Dover, 1992). Sir Isaac Newton published his observations of the colours of soap bubbles in *Opticks* (London: G. Bell and Sons, 1730).

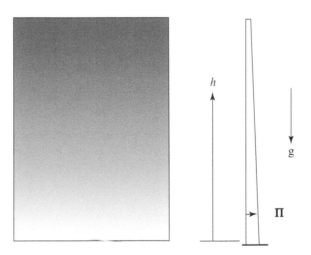

Figure 10.7   Schematic diagram of the effect of drainage under gravity on the thickness of a large vertical soap film.

This experimental set-up has been used to measure the interaction force between surfactant surfaces, as a function of separation distance or film thickness. These forces are important in stabilising surfactant lamellar phases and in cell-cell interactions, as well as in colloidal interactions generally.

FROTH FLOTATION

Materials mined from the earth's crust are usually highly heterogeneous mixtures of amorphous and crystalline solid phases. Crushing and grinding operations are employed to liberate individual pure grains in the 10–50 μm size range. One of the most widely used (10^9 tons per year) processes for separating the required mineral from the gangue (unwanted material e.g. quartz) exploits the wetting properties of the surface of the grains in a froth flotation process. This cheap and relatively simple process is based on the observation that hydrophobic (or non-wetting) particles dispersed in water easily attach to air bubbles which carry them upwards to the top of the flotation chamber where they can be collected. Hydrophilic particles do not attach to the bubbles and remain dispersed in bulk solution. This important industrial process was invented in Australia by Charles Potter a Melbourne brewer in 1901. The first commercial process was set up in Broken Hill.

The common gangue material quartz (silica) is naturally hydrophilic and can be easily separated in this way from hydrophobic materials such as talc, molybdenite, metal sulphides and some types of coal. Minerals which are hydrophilic can usually be made hydrophobic by adding surfactant (referred to as an activator) to the solution which selectively adsorbs on the required grains. For example, cationic surfactants (e.g. CTAB) will adsorb onto most negatively charged surfaces whereas anionic surfactants (e.g. SDS) will not. Optimum flotation conditions are usually obtained by experiment using a model test cell called a Hallimond tube. In addition to activator compounds, frothers which are also surfactants are added to stabilise the foam produced at the top of the flotation chamber. Mixtures of non-ionic and ionic surfactant molecules make the best frothers. As examples of the remarkable efficiency of the process, only 45 g of collector and 35 g of frother are required to float 1 ton of quartz and only 30 g of collector will separate 3 tons of sulfide ore.

The flotation process is used in the early separation stage for obtaining pure metals from Cu, Pb, Zn, Co, Ni, Mo, Au and Sb ores, whether sulphides, oxides or carbonates. It is also used to concentrate CaF_2, $BaSO_4$, NaCl, KCl, S, alumina, silica and clay. Ground coal is also treated to remove ash-producing shale, rock and metal sulphides which cause air pollution by SO_2 during combustion. In recycling processes, ink can be removed from paper and metallic silver from photographic residues using flotation. It is even used for removing earth from vegetables during cleaning.

THE LANGMUIR TROUGH

Surfactant molecules will adsorb onto a wide range of solid substrates from aqueous solution. The amount and type of adsorption depends on the solution concentration, the nature of the surfactant and the substrate. Generally, ionic surfactants adsorb on oppositely charged surfaces and non-ionic ones adsorb on most surfaces. Often, monolayer and bilayer adsorption is observed, as shown below. In this example, the cationic surfactant CTAB adsorbs on mica (a negatively charged aluminosilicate) with a well-defined step adsorption isotherm:

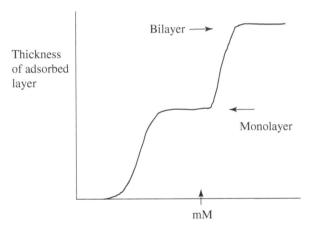

Figure 10.8   Schematic diagram of a simple monolayer and bilayer surfactant adsorption from an aqueous solution.

A monolayer usually adsorbs well below the cmc, whereas a bilayer forms only at the cmc (in this case around 1mM at 30 °C). Since the mica substrate is hydrophilic, we would expect the water contact angle to be a maximum for the hydrophobic monolayer and then fall when the bilayer adsorbs, as has been reported. These observations correlate well with the flotation behaviour of the mineral. Clearly, surfactant adsorption can completely alter the wetting properties of a substrate surface and can is also used to control the coagulation of sols via their effect on the particle's charge.

In the case of adsorption from solution, the surfactant layers are in equilibrium with the solution and well desorb on dilution. However, it would be very useful to produce adsorbed layers in both air and water which will remain adsorbed. This can be achieved using the Langmuir-Blodgett deposition technique. The technique is based on the observation that if a surfactant, which is insoluble in water, is dissolved in a volatile, non-aqueous solvent and then spread on water, an insoluble monolayer of orientated surfactant molecules will remain at the air/solution interface. The effect of the spreading surfactant and its surface film pressure can be dramatically demonstrated by spreading hydrophobic talc powder on a clean water surface and then placing a droplet of surfactant solution gently in the middle. The talc is rapidly displaced in all directions from the centre by the spreading surfactant. This is the basis of the camphor duck toy, which swims around a water bath via dissolution and surface spreading of a small piece of camphor attached to its tail.

The behaviour of spreading films on water and the old observation of 'pouring oil on troubled waters' has a long history. Pliny the Elder, who died in the Mt Vesuvius eruption of AD 79, refers to spreading of oil on water, as does the Venerable Bede (673–735). More recently, Benjamin Franklin did his experiments, on the area of water which could be stilled by a spreading film, on Clapham Common, and his results were transmitted by his friend William Brownrigg to the Royal Society, which promptly published it (with slight modifications) in *Philosophical Transactions of the Royal Society*, Vol. LXIV, p. 447 (1774). Later, Lord Rayleigh made the important assumption that the spreading film was monomolecular to make the first estimate of molecular size. His estimate was actually quite close to the thickness of a film of oleic acid, the main component in olive oil. Lord Rayleigh's experiments were reported in *Proceedings of the Royal Society*, Vol. XLVII, p. 364 (March 1890).

Spreading of an insoluble (or temporarily insoluble) surfactant monolayer effectively produces a two-dimensional surface phase. This thin molecular layer exerts a lateral film pressure, which can be simply demonstrated by covering a water surface with a uniform layer of finely divided hydrophobic talc and placing a droplet of surfactant solution (0.003 M CTAB solution) in its centre. The effect of the film pressure of the spreading surfactant is dramatic, as seen on the following photos.

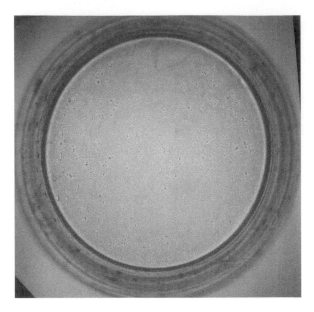

Figure 10.9 Photograph of hydrophobic powdered talc spread uniformly on the surface of a dish of clean water.

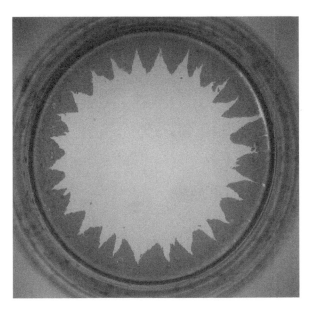

Figure 10.10 Instantaneous removal of the talc in the centre caused by the addition of a droplet of soap solution.

Talc symmetrically removed from the centre by the application of one droplet of 0.003 M CTAB solution. The process occurs rapidly (in less than 0.1 s) due to the surfactant's rapid rate of surface diffusion. The hydrophobic talc is compressed against the glass walls of the crystallizing dish. Placing more droplets at the glass edges displaces the talc layers, which move like tectonic plates to produce scaled maps of the world, à la Slartibartfast.

We can study these lateral surface forces by measuring their action on a freely moveable beam separating the (insoluble) surfactant coated surface from the pure water surface, as illustrated below:

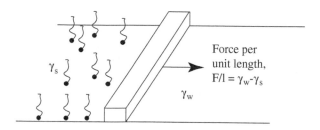

Figure 10.11   Schematic diagram of the forces acting on a (Teflon) beam separating a pure water surface from a surfactant monolayer coated surface.

That the force per unit length (F/l) acting on the barrier is given by $(\gamma_w - \gamma_s)$ can be easily shown by consideration of the free energy change on allowing the barrier to move an infinitesimal distance to the right, thus:

$$dG = +ldx\gamma_s - ldx\gamma_w$$

and the force acting on the barrier must be given by the gradient in free energy of the system:

$$F = -dG/dx \text{ and hence: } F/l = \gamma_w - \gamma_s$$

The force per unit length (F/l) generated by the surfactant film is called the surface film pressure Π_F. In the Langmuir trough device, illustrated below, the density and hence the pressure in the film can be varied via a movable barrier:

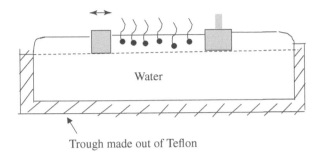

Water

Trough made out of Teflon

Figure 10.12 Schematic sectional diagram of a Langmuir trough showing a surface trapped film of insoluble surfactant molecules.

The first description of a Langmuir trough appears to have been given by Langmuir in *Journal of the American Chemical Society*, Vol.39, p. 1848. Agnes Pockels carried out many experiments which were reported in *Nature*, Vol. 43, p. 437 (1891): a remarkable achievement for a woman of the time, working from home. A schematic diagram of a typical experimental set-up is given below. In the case illustrated, the surface tensions are measured using the rod-in-free surface technique, and a motor is used to move the Teflon boom.

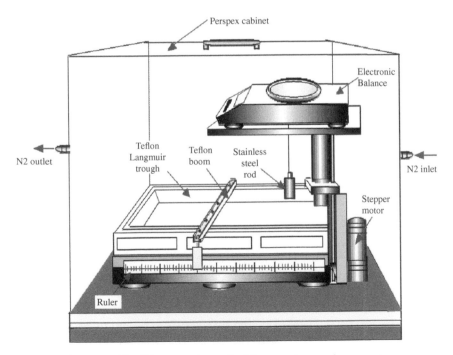

Figure 10.13 Diagram of a typical Langmuir trough apparatus.

If a solid substrate is withdrawn vertically through the surfactant side of the Langmuir trough, a uniform monolayer can be transferred to the solid, as illustrated in the following figure:

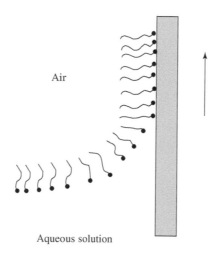

Figure 10.14   Langmuir-Blodgett coating of a surfactant monolayer.

Since the density of the monolayer can be varied using the applied surface pressure (via movement of the boom), a wide variety of monolayer conformations can be deposited. Because the surfactant is insoluble, the layer will be stable in both air and aqueous solutions. A second bilayer can be adsorbed on the plate if it is re-immersed in the trough. The layers adsorbed using this technique are called Langmuir-Blodgett (LB) layers. An example of a close-packed surfactant LB coating on a smooth polished substrate, as imaged by an atomic force microscope, is given below:

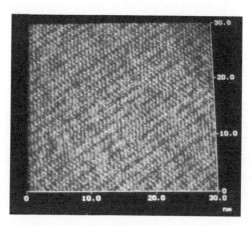

Figure 10.15   Atomic force microscope image of a Langmuir-Blodgett surfactant monolayer coated on a smooth solid substrate.

Using special surfactants, the coated layers can be polymerised to increase their strength (e.g. C_{11}-C≡C-C≡C-C_7COOH polymerises on exposure to UV irradiation). These monolayers improve the characteristics of photo-semiconductors. A power increase of 60% can be achieved by the deposition of only two monolayers. LED devices can be improved in efficiency by an order of magnitude by depositing 8 surfactant layers, apparently because of a tunnelling mechanism. The alternative method of vacuum deposition of thin layers causes damage to the device because of heating at the surface during evaporation and, in addition, does not produce such uniform layers. LB layers are now also being used for photoresist masking on silicon. The resolution at present is about a micron, but by using LB layers, it could be reduced to 0.1 micron. Finally, LB films also act as an effective lubrication layer between solids.

Since the total number of molecules of surfactant added to the Langmuir trough is known (from the molecular weight), the area per molecule is also known and can be varied simply by moving the boom. The relation between the film pressure and the area per molecule can, therefore, be measured. This is in fact a very elegant method for the study of molecular films. The precise isotherm is characteristic of the surfactant, but the general features often observed are shown below:

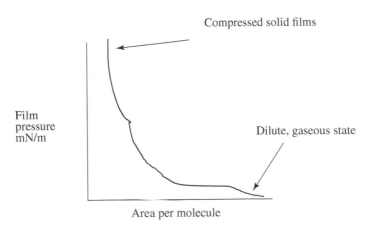

Figure 10.16   Typical film pressure isotherm for a surfactant monolayer.

The various regions on the isotherm are determined by the lateral interaction between the surfactant molecules within the surface phase. In the dilute, gaseous state, the molecules can be considered to be negligible in size and non-interacting. Under these conditions, the isotherms obey an ideal, two-dimensional gas equation of the form $\pi A = kT$. As the pressure is increased, a point is reached (at about 8 nm^2 for myristic acid) where the attractive forces between the hydrocarbon tails cause a condensation process, analogous to liquid condensation from the vapour phase. However, at the end of this process (at about 0.5 nm^2 in the example given) the surface layer is not completely condensed because of the strong, relatively long-range electrostatic repulsion between the ionic head-groups. This head-group repulsion keeps the surface layer fluid, whilst the attractive van der Waals forces between the hydrocarbon chains keep the film coherent. In this state, a modified ('real') gas equation can be used to describe the isotherm of the form:

$$(\pi - \pi_0)(A - A_0) = kT$$

where π_0 and A_0 are correction terms related to the attractive and repulsive forces between the molecules.

At still higher pressures, the film becomes completely packed (and will eventually buckle), and the limiting area corresponds to the cross-sectional packing area of the surfactant molecule. This region is also interesting because it demonstrates that compressed surface films will respond to even small increases in surface area, e.g. by stretching a surface through mechanical vibrations, with a large increase in surface energy of the entire film. This behaviour generates a surface elasticity which is important in giving mechanical stability to soap films (discussed earlier) and is also the cause of the 'oil on troubled waters' phenomenon, observed for more than two millennia. For every breath we take, lung surfactants (lipids) are similarly compressed and released as an important part of our physiological respiratory processes related to the control of surface tension.

For more information, see the standard text: G. L Gaines, *Insoluble Monolayers at Liquid-Gas Interfaces* (New York: Interscience, 1966).

Laboratory Experiment   Flotation of powdered silica

Introduction

Froth flotation is one of the simplest and most widely used separation techniques for mineral ores. Crushing and grinding of highly heterogeneous rocks liberates individual pure grains in the 10–50 μm size range. These particles can often be readily separated using differences in their surface properties. When one of the components is hydrophobic and the other hydrophilic, gas bubbles passed through a stirred aqueous suspension carry the hydrophobic particles to the top of the vessel, where they can be collected. The wetting properties of the surface of the pure mineral determines the flotation efficiency. Thus, if water has a high contact angle on the mineral its flotation efficiency will also be high.

Often the minerals we want to float are hydrophilic, and surfactants (called 'collectors') are added which, at a specific concentration, adsorb onto the particle surface, making it hydrophobic and hence floatable. In a mixture of hydrophilic minerals, optimum flotation will occur where one of the minerals adsorbs collector but not the others.

In industrial processes the flotation cells have, of course, a very large capacity. However, the efficiency of the process can be determined on a much smaller scale using a Hallimond apparatus (see the figure later). In this experiment we will use this apparatus to measure the flotation efficiency of hydrophilic Ballotini beads (~50 μm diameter) over a range of CTAB (cetyltrimethylammonium bromide) concentrations. The glass beads (when clean) are naturally hydrophilic because of a high density of surface silanol groups which hydrogen-bond strongly with water. Adsorption of a monolayer of CTAB makes the surface hydrophobic, but at higher CTAB concentrations a bilayer adsorbs, which makes the surface hydrophilic again. It is the solution concentration that controls this adsorption and hence the flotation efficiency of glass.

Experimental procedures

The Ballotini glass beads (~50 μm diameter) used in these experiments should be cleaned in dilute NaOH solution, washed thoroughly with distilled water and then completely dried. The surface of the powder should then be hydrophilic, with a high density of silanol groups (about 1 per 25 Å^2) which makes the particles hydrophilic.

A glass Hallimond tube apparatus typically used to study flotation in the laboratory is illustrated in the following figure.

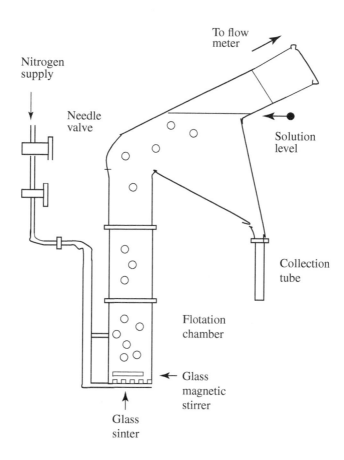

Instructions

(1) Dismantle and clean the flotation apparatus by scrubbing with ethanol, followed by rinsing with distilled water.

(2) Put the apparatus together as shown in the diagram but leave off the top. Close the tap connecting the gas line to the flotation chamber.

(3) Fill with a known volume of distilled water to just below the solution level indicated in the figure. The top of the apparatus must be connected to a suitable low rate flowmeter. (A soap bubble column is adequate.)

(4) Adjust the needle valve until the rate of N_2 gas flow rate is about 50 mL/min. Note this rate.

(5) Add distilled water (whilst the gas is still flowing) to bring to the solution level to that indicated by an arrow in the figure, and note the total solution volume. (At this level particles carried up to the surface will be collected and will not return to the flotation chamber once the bubbles rupture at the solution surface.)

(6) Switch off (or bypass) the gas flow such that the flow rate remains the same on reopening the flow. Consistency of the flow rate is necessary to obtain accurate comparisons of flotation efficiency. Check this by remeasuring the flowrate.

(7) Disconnect the N_2 line (at the needle valve on the gas cylinder) and pour out the distilled water.

(8) Accurately weigh about 1 gm of clean, dry Ballotini powder and place into the flotation chamber, half fill with a known volume of distilled water and drop in the glass magnetic stirrer supplied. (Be careful as the latter is easily broken.) Accurately weigh the dry collection tube and then fit onto the apparatus.

(9) Stir the powder and solution for five minutes to 'condition' the surface, then allow it to settle.

(10) Reconnect the rest of the apparatus to the flotation chamber and add more distilled water to bring the mixture to the already determined total solution volume. Connect to the flow meter and begin stirring the flotation chamber.

(11) Vent the N_2 to atmosphere for a few seconds, if using a bypass system, to remove pressure build-up in the line prior to each run.

(12) Allow the N_2 gas to flow for precisely 30 seconds, making sure that the flow rate remains the same as before. Switch off the magnetic stirrer and allow five minutes to settle.

(13) Pour off the excess solution from the apparatus and remove the collection tube. Allow the powder to settle in the tube, and then pour off as much excess solution as possible without losing any of the powder and dry at about 120 °C for at least 20 minutes in an oven (loosely cover to prevent dust settling on the powder during drying) or leave overnight. Reweigh the collection tube.

(14) Calculate the percentage flotation efficiency, that is, % collected/original weight.

(15) Repeat and take the average value.

(16) Follow the procedure described above using 1 gm portions in 10^{-7}, 10^{-6}, 10^{-5}, 10^{-4} and 10^{-3} M CTAB solutions. Addition of CTAB may cause extensive frothing. Since the powder will be

retained in this foam, it must be broken up either mechanically using a Pasteur pipette or by the addition of a small amount of ethanol. The powder may also stick to the walls of the apparatus above the collection tube. This powder should be dislodged by gentle tapping on the walls of the apparatus. Calculate the mean flotation efficiency at each concentration and graph the results.

Contact angles

Clean a soda glass plate by washing in 10 % NaOH (with care) followed by rinsing in double-distilled water. Blow dry with N_2 and observe the behaviour of a droplet of clean water on the plate. Blow dry again, place droplets of the various CTAB solutions used on the clean plate and observe the wetting behaviour with CTAB concentration.

FOR CONSIDERATION/TYPICAL QUESTIONS

(1) Explain the flotation recovery results obtained with CTAB.
(2) What are the major differences between this experiment and a typical industrial flotation process (other than size)?
(3) Why did ethanol break up the surfactant foam?

11

Bubble Column Evaporators

The bubble column evaporator or BCE process is based on the continuous production of gas bubbles in an aqueous column, as a source of high gas-water interface, which provides an efficient process for the controlled transfer of heat and mass.

THE BUBBLE COLUMN EVAPORATOR PROCESS

Although the bubbling process itself is both simple to use and apply, understanding the fundamental physical and chemical principles involved is surprisingly difficult, and there are many issues yet to be properly explained. Recently the process has been used to develop new methods for the precise determination of enthalpies of vaporisation (ΔH_{vap}) of concentrated salt solutions, as an evaporative cooling system, a sub-boiling thermal desalination unit, for sub-boiling thermal sterilization, for low-temperature thermal decomposition of different solutes in aqueous solution and for the inhibition of particle precipitation in supersaturated solutions. Very different results have been obtained using nitrogen, oxygen, carbon dioxide and helium inlet gases. These novel applications can be very useful in many industrial practices, such as desalination, water/wastewater treatment, thermolysis of ammonium bicarbonate (NH_4HCO_3) for the regeneration in forward osmosis and refrigeration-related industries. The background theories and models use to explain the BCE process are also reviewed, and this fundamental knowledge is applied to the design of BCE systems and to explain recently explored

Applied Colloid and Surface Chemistry, Second Edition. Richard M. Pashley and Marilyn E. Karaman.
© 2021 John Wiley & Sons Ltd. Published 2021 by John Wiley & Sons Ltd.
Companion website: www.wiley.com/go/pashley/appliedcolloid2e

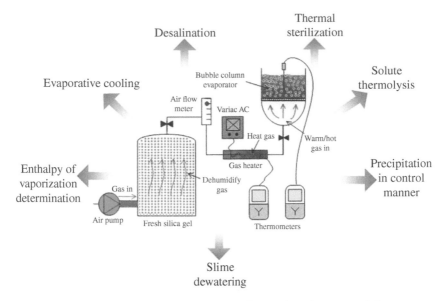

Figure 11.1 Diagrammatic summary of several applications of the BCE.

applications, as well as potential improvements. There are many prospective applications of the BCE process, and some are illustrated in the diagram.

Bubble column evaporators or BCEs are simple devices that have emerged as facilitators of powerful new technologies for aqueous systems in:

- Desalination (seawater and groundwater)
- Water sterilization (recycled water with no surviving pathogens including viruses)
- Thermolysis of solutes in aqueous solution (destruction of drugs, facilitation of high temperature reactions)
- Inhibition of salt precipitation in concentrated salt solutions
- Evaporative air conditioning systems for buildings
- Slime dewatering

The first three are now proven processes and cheap at industrial scales. They represent encouraging progress in the search for clean water, arguably the biggest problem facing the world.

Heat transfer may also be accomplished directly by mixing the solution and the heating fluid (for example, water and hot air bubbles) to give so-called direct-contact evaporators. This concept of heat transfer via hot bubbles was first demonstrated by Collier in a patent, published in 1887, and the first commercial plant was installed in the United States in the early 20th century. The advantages of bubble columns using direct-contact heat transfer, compared to other multiphase reactors, are several: (a) less maintenance is required due to the absence of moving parts, (b) higher effective interfacial areas and overall mass transfer coefficients can be achieved, (c) higher heat transfer rates per unit volume of the reactors can be attained, (d) solids can be handled without any erosion or plugging problems and (e) less floor space is occupied and bubble column reactors are less costly. Finally, and a glaringly obvious technology still to be exploited, high-temperature reactions can be carried out at the surface of bubbles whilst maintaining a relatively low temperature in the liquid column.

The BCE is easy to use. But a theoretical understanding of the processes involved is still embryonic. It becomes more difficult to understand when a new phenomenon is thrown into the mix. This is the phenomenon of bubble coalescence inhibition with addition of salt. This was first used experimentally by Russian mineral coal flotation engineers more than 100 years ago. The addition of sufficient salt inhibits bubble-bubble fusion, produces smaller bubbles and improves the efficiency of flotation. More recently, aqueous bubble column evaporators have been used for a range of new applications based largely on the unexpected effects of many salts and sugars on inhibiting bubble-bubble coalescence in water, in combination with limited bubble rise rates and rapid water vapour uptake into the bubbles.

The visual difference between a bubble column containing distilled water and seawater is caused by the inhibition of coalescence by the latter, as seen in the following photographs where only salt levels were varied.

Even more remarkable is the fact that the ionic strength of blood is exactly the same critical concentration where we see maximum bubble coalescence inhibition. There must be good physiological reasons for this which must have important links to our evolution from the sea.

The observation that the addition of common salt inhibits bubble coalescence presents us with a useful process (e.g. it is used in froth flotation) but

Pure water   Salt solution

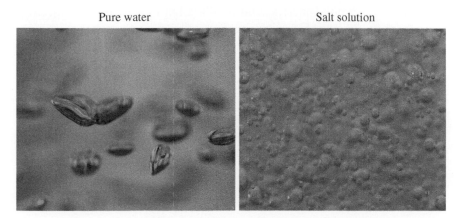

Figure 11.2   The effect of added salt on bubble coalescence.

unfortunately also presents a problem for our understanding as to why this actually occurs. For example, bubbles are negatively charged in water due to the relative excess surface adsorption of OH- ions (i.e. inferred from zeta potential measurements), and so the electrostatic double layer repulsion between two colliding bubbles should actually be reduced by adding salt. Remember that the van der Waals force between two symmetrical air bubbles will be attractive and little affected by salt. Further to this, the addition of salts to water increases its surface tension and so bubble coalescence and consequent reduction in surface area will reduce the overall surface free energy, giving an increased driving force for coalescence. The opposite behaviour observed remains a significant issue which has yet to be resolved! Even so, bubble coalescence behaviour is vital in water-based cleaning processes, ore flotation, food processing, gas–oil separations, absorption and distillation. It is possible that we fail to understand these bubble coalescence effects completely because the cause is more closely related to dynamic effects occurring during bubble-bubble collisions. For example, two bubbles slowly forced together in water or salt solution always coalesce. Clearly, this is one area requiring further research effort. We are, however, able to produce columns of flowing bubbles of controllable bubble size, and understanding their behaviour opens up many new processes and important commercial applications.

BUBBLE WATER VAPOUR EQUILIBRATION

In the BCE process, control of the size of the bubbles produced is very important. This is because the finest bubbles produce the largest active surface area or water/gas interfacial area, but unfortunately small bubbles rise very slowly and so are not useful for efficient mass and energy transfer processes. It turns out that air bubbles in water larger than about 1–3 mm diameter become non-spherical and oscillate both in shape and trajectory, reducing their rise rate. Using a suitable glass sinter, it is easy to produce a continuous flow of bubbles of this optimum size, especially using added salt to inhibit bubble coalescence. A high-density bubble column can be produced this way, as shown in the following figure.

It is remarkable that water vapour saturation within these rising bubbles is attained rapidly, within a few tenths of a second, because of slight oscillations in bubble shape and the circulatory fluid flow induced inside the bubbles due to shear forces generated at the surface of rising bubbles. The water vapour transfer is much faster that that expected for quiescent diffusion, which would require several seconds to reach equilibrium according to Fick's law. This rapid vapour transfer must correspond to a similarly fast

Figure 11.3 High-density (non-boiling) bubble column formed to desalinate seawater.

rate of transfer of heat to the surrounding column solution. These factors form the basis for several recent applications of the BCE process.

BUBBLE RISE VELOCITY

Bubble rise behaviour in water, even for single isolated air bubbles, is surprisingly complex and depends on bubble diameter, sphericity and water purity. The presence of many other bubbles within a BCE makes this situation even more complex, and this has not been well studied. It was demonstrated by Leifer et al. in 2000 (in the *Journal of Atmospheric and Oceanic Technology*) that the motion of intermediate (single, isolated) bubbles ranging from 1–3 mm in diameter is produced by the combination of two oscillation types, trajectory oscillations (zig-zag or helical) and shape or deformation oscillations (ellipsoidal). These gas bubbles actually rise at a limited rate of between about 15 and 35 cms^{-1} in quiescent water because they undergo trajectory and shape oscillations which reduce their rise rate. Hence, looking at the data given below it is clear that in water air bubbles in the 1–3 mm range are the smallest that will rise at the fastest rate.

As can be seen from the graphs, using Stokes's law for spherical objects moving under high Reynolds number and with zero slip boundary condition gives rise to rates substantially higher than those observed for air bubbles in water. Unfortunately, the addition of slip boundary conditions gives even higher rise rates, such as those obtained using the Hadamard-Rybczynski (H-R) equation. It seems that the real behaviour of larger bubbles is dominated by their deviation from spherical shape.

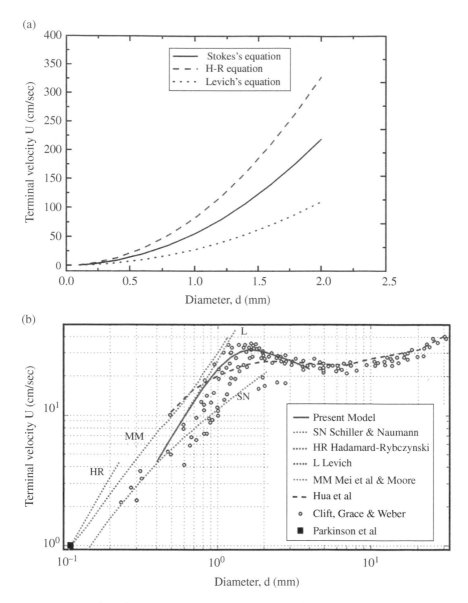

Figure 11.4 The relationship between rise velocity of isolated bubbles and bubble diameter using typical basic equations is given in (a), and models are compared with experimental results in (b). [Published by Klaseboer E. et al., "BEM simulations of potential flow with viscous effects as applied to a rising bubble," *Eng. Anal. Bound. Elem.*, 35(3): 489–494, 2011.]

Surfactants can be added to modify the surface of bubbles predominately through the adsorption of a monolayer surface coating. This produces a more rigid interface and so enhances fluid drag, so the rise velocity in these solutions is less than that for clean bubbles of the same size. The complex behaviour of bubble rise rates for isolated bubbles in the diameter range >1 mm will be further complicated by multiple collisions within a densely packed bubble column, where coalescence is inhibited by the presence of added salts.

THERMAL ENERGY BALANCE IN THE BCE

Simple experiments on bubble columns tells us that the continuous flow of dry air into a water column causes a significant evaporative cooling effect. For example, dry room-temperature air flow will cool a water column down to a steady state temperature of about 10–12 °C. The steady state thermal energy balance within a BCE can be used to explain the process whereby the heat supplied from the entering (relatively) warm bubbles (per unit volume of gas leaving the column) will be balanced by the heat required for water vaporisation to reach the equilibrium water vapour pressure within these bubbles. A steady-state volumetric balance within a bubble column has been derived and has been used for the determination of the enthalpy of vaporisation (ΔH_{vap}) of concentrated salt solutions. This equation is based on a simple dynamic thermal energy balance and is described by the following equation:

$$\left[\Delta T \times C_p\left(T_e\right)\right] + \Delta P = \rho_v\left(T_e\right) \times \Delta H_{vap}\left(T_e\right) \qquad (11.1)$$

where $C_p(T_e)$ is the specific heat of the gas flowing into the bubble column at constant pressure; T_e is the steady state temperature near the top of the column; ρ_v is the water vapour density at T_e, which can be calculated from the water vapour pressure of salt solutions at the steady state temperature, using the ideal gas equation; ΔT is the temperature difference between the gas entering and leaving the column; ΔP is the hydrostatic differential pressure between the gas inlet into the sinter and atmospheric pressure at the top of the column, which represents the work done by the gas flowing into the base of the column until it is released from the solution.

This equation was first presented in 2009 by Francis and Pashley (in the *Journal of Physical Chemistry*) for low vapour pressure aqueous solutions, that is, at low column temperatures of about 283 K, and was successfully used to measure the enthalpy of vapourisation of concentrated salt solutions. The apparatus used for these measurements is described below.

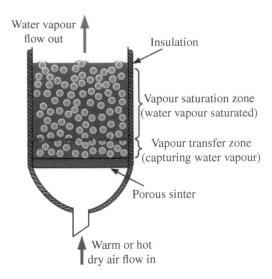

Water vapour
flow out

Insulation

Vapour saturation zone
(water vapour saturated)

Vapour transfer zone
(capturing water vapour)

Porous sinter

Warm or hot
dry air flow in

Figure 11.5   Schematic diagram of a basic BCE apparatus.

It should be noted that since bubbles reach vapour and temperature equilibrium within a few tenths of a second, the column height becomes important. Hence, the column height should be selected to correspond to one where this equilibrium is just attained (i.e. it needs to be at least 10 cm in height).

Bubble coalescence inhibition and rapid vapour transfer both offer a novel application of the BCE process for the precise measurement of ΔH_{vap} values for concentrated salt solutions where the vapour pressure values of salt solutions are known. ΔH_{vap} values can be obtained from the volumetric energy balance within the column, operating under steady state conditions, as defined by equation (11.1), at ambient temperatures. Using this method, ΔH_{vap} values of various salt solutions at different concentrations can be obtained with accuracies, on average, within around 0.5–1.0% compared to the literature values.

FURTHER APPLICATIONS OF THE BUBBLE COLUMN EVAPORATOR (BCE)

The gas-liquid, direct-contact bubble column evaporator is characterized by a continuously replenished, high gas-liquid interfacial area, which subsequently offers higher mass and heat transfer coefficients due to the non-isothermal (i.e. localized evaporation) nature of BCE and can be applied in many large-scale industrial applications. The bubble column evaporator (BCE) currently has the following range of applications in aqueous systems:

BCE FOR EVAPORATIVE COOLING

The steady-state operating temperature of an aqueous solution in a bubble column can be calculated using the volumetric energy balance equation (11.1). The relation between inlet gas temperature and column temperature, using known ΔH_{vap} values and vapour pressure values for salt solutions or pure water, can be calculated for pure water and salt solutions. The steady-state column solution temperature within the BCE is a function of the temperature of the inlet gas. This cooling effect has led to the suggestion that the BCE process could be used as a simple evaporative cooling system for buildings. The presence of high salt levels, which are retained in the solution, will also prevent the growth of dangerous microorganisms, which are a problem with open or trickle-bed evaporative cooling systems.

When the BCE process is used with some types of salt solutions, which have coalescence inhibition effects on the bubbles, it will produce a high volume fraction of small bubbles which will enhance the water vapour transfer into the bubbles. Thus, a solution with dissolved salt (NaCl) at 0.5 M has a more efficient evaporative cooling effect and hence halved the time for the column to reach steady state conditions, compared with pure water. As an example, a continuous flow of dry inlet air at a temperature of 50 °C will cool the 0.5 m NaCl column solution to less than 20 °C.

SEAWATER DESALINATION USING THE BUBBLE COLUMN EVAPORATOR

Water vapour can be captured and transported using a simple BCE system operated at temperatures well below the solution's boiling point. The inhibition of bubble coalescence in salt solutions enables the design of a bubble column with a high-volume fraction of small air bubbles, continuously colliding but not merging. This produces a uniform and efficient exchange of water vapour into the bubbles, which together with the high bubble rising velocity, due to its shape and trajectory-based oscillations, allows water vapour to be rapidly absorbed into bubbles, condensed and then collected as pure drinking water. The BCE method for seawater desalination was examined

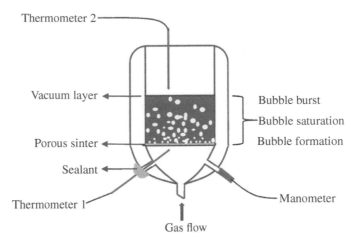

Figure 11.6 Schematic diagram of a monitored BCE apparatus used for the study of the thermal efficiency of desalination.

and patented in 2013. The process is, of course, a reduced version of the natural phenomenon in which air is used as a carrier gas for desalting seawater through the rain cycle. However, the BCE process is based on the unexpected property of seawater inhibiting air bubble coalescence because this facilitates a high packing volume of air bubbles, which are persistently colliding but are prevented from coalescing by the presence of salts. In addition, the bubbles so produced in the 1–3 mm diameter range are ideally suited for rapid water vapour uptake. These factors form the basis of this enhanced process for the desalination of seawater. A schematic diagram of the bubble column desalination process is shown below.

ENHANCED SUPERSATURATED BUBBLE COLUMN DESALINATION

The addition of a non-volatile, non-ionic surfactant (such as octa-ethylene glycol mono-dodecyl ether, $C_{12}EO_8$) increases the water vapour carryover expected from the BCE solution vapour pressure for any given air volume passed. It seems likely that the use of the non-ionic surfactant provides a monolayer coating at the surface of the rising bubbles and this packed monolayer of surfactant molecules allows water vapour transport into the bubbles but inhibits this vapour from recondensing on the interior of the now hydrophobic walls of the bubbles. Hence, the surfactant layer acts like a 'surface molecular diode', which facilitates water vapour molecular transport in one direction. This supersaturation of the air bubbles then produces increased water vapour carryover and so more efficient desalination.

ENHANCED BUBBLE COLUMN DESALINATION USING HELIUM AS A CARRIER GAS

Using dry heated helium as the inlet gas in a BCE system can also increase the water vapour capture efficiency and, in this case, by more than 3.3 times higher compared with the expected equilibrium vapour collection for dry air. This is a significant increase because commercial desalination processes, typically based on reverse osmosis (RO), are still energy costly, and so an improved efficiency in such a simple non-boiling process might offer substantial benefits in many parts of the world.

It is therefore important to ask: Why does helium gas behave in this way? The small size of helium atoms combined with their heated state might facilitate their flow into the water layers surrounding the rising bubbles. This may cause a proportion of the hydrogen bonding network within these water molecules to break, which would enhance water vapour collection in the bubbles and hence improve desalination efficiency. For example, a decrease in hydrogen bonding in liquid water of 3.6 to 3.2, due to a modest increase in temperature, from 0 to 70 °C, corresponds to a decrease in ΔH_v of about 3 kJ/mol, and this effect will cause an increase in water vapour pressure. Could this effect produce the significant increase in water vapour pressure?

From the standard equations of equilibrium we have:

$$\Delta G = \Delta H - T\Delta S \tag{11.2}$$

and

$$\Delta G^0 = -RTlnK_{eq} \leftrightarrow K_{eq} = \exp\left(-\frac{\Delta G^0}{RT}\right) \tag{11.3}$$

It can be assumed that the entropy difference of the conversion of water to gas (liquid $\leftrightarrow$ gas) will be constant for either air or helium. So any difference in vaporization entropy change for the two gases will be insignificant, that is: $d(\Delta S) = 0$ and so Equation (11.3) becomes:

$$d(\Delta G) = d(\Delta H_v) \tag{11.4}$$

Given this and the water density ratio of helium compared to air $\left(K_{eq} = \dfrac{\rho_{He}^{w}}{\rho_{air}^{w}} = 3.3\right)$, Equation (11.4) becomes:

$$K_{eq} = \frac{\rho_{He}^{w}}{\rho_{air}^{w}} = 3.3 = \exp\left(-\frac{d(\Delta H_v)}{RT}\right) \tag{11.5}$$

In this equation, R is the gas constant and T is the average column temperature, and hence $d(\Delta H_v)$ is calculated to be about –3 kJ mol^{-1}. The expected ΔH_v values for different column temperatures is, on average, about 43.4 kJ mol^{-1}, so with helium experiments, a 7% reduction of ΔH_v is sufficient to explain the enhanced water vapour carryover observed.

A schematic diagram of the combined effects for heated helium gas sparging and the use of added surfactants is shown in the following figure. This gives a description of the mass transfer between helium atoms and water molecules in the BCE process. In the figure, (A) is an illustration of helium bubbles coated with surfactant with hydrophilic heads facing the water-soluble phase and hydrophobic tails orientating the gas phase, (B) is an illustration of the 'molecule diode' effect caused by adsorbed surfactant which facilitates water vapour entry and also inhibits the exit of water molecules and (C) illustrates the resulting saturated water vapour bubbles.

The main advantages of the BCE desalination system are its simplicity, resilience to feed water purity and the fact that it is a continuous and controlled non-boiling process. These are clear advantages over the two most common seawater desalination processes currently used, that is, seawater reverse osmosis (SWRO) and thermal desalination (such as MSF). There is little room left for improvement in SWRO, but thermal desalination methods can still be substantially improved. Methods such as BCE, which represent low capital investment, do not rely on rare materials or complex manufacturing and ready use of waste heat and sustainable energy sources, such as wind power, and offer fewer constraints than the other common processes.

WATER STERILIZATION USING BCE

The high heat transfer coefficients created within the BCE system can be used to thermally destroy biological organisms in water well below the boiling point, and it has recently been established that water sterilization can be produced via the transitory collisions between biological species and hot rising gas bubbles, even though the column temperature remains low. Hot

A. Surfactant surrounded dry helium bubble

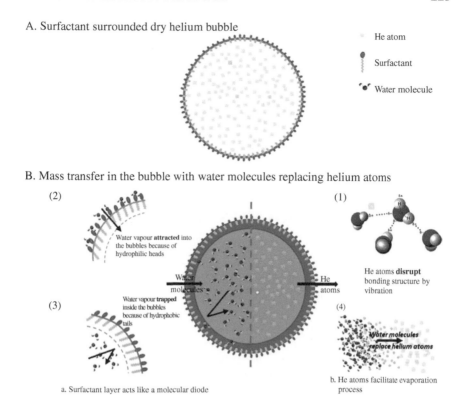

B. Mass transfer in the bubble with water molecules replacing helium atoms

C. Water molecules saturated bubbles

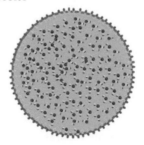

Figure 11.7 Schematic diagram of a proposed mechanism for helium-catalysed BCE desalination.

gas bubbles up to 250 °C can be passed into a water column via a glass sinter with 40–100 μm pores. The effects of exposure on sterilizing water show that only 2 minutes of flow of 250 °C air were required to destroy almost all of the coliforms present in contaminated wastewater.

Once again, the presence of added salt that inhibits bubble coalescence in the solution serves to preserve finer bubbles of the heated gas, enhancing the number of bubble collisions with the biological species by ensuring a higher gas-liquid contact surface area and higher surface area per gas volume which leads to improved sterilization rates.

Recently, it has been discovered that gas bubbles of CO_2 can be very effective at killing standard waterborne viruses, even at 1 atm and at room temperature, although more effective killing rates were obtained using heated CO_2 bubbles. It has recently been suggested that this gas, combined with O_2 and warm water vapour, could be used for the destruction of the COVID-19 virus present in the lungs.

THERMOLYSIS OF SOLUTES IN AQUEOUS SOLUTION

Successful thermal sterilization studies led to the suggestion that this process could also facilitate thermal decomposition of some solutes in aqueous solutions, even at lower solution temperatures and at a faster rate than is normally produced via the direct heating of a bulk solution. The important salt ammonium bicarbonate (NH_4HCO_3) in aqueous solution is one example of a salt which can be thermally decomposed using the BCE process with hot air. This salt has been used for important applications, such as a draw solution in forward osmosis (FO), and more recently in the regeneration of ion exchange resins.

This salt is important because it can be decomposed at relatively low temperatures (> 60 °C) into gaseous products which can be captured and used to reform the salt. The thermal decomposition reaction is given by:

$$NH_4HCO_{3(aq)} \overset{\Delta}{\rightleftharpoons} NH_{3(g)} + CO_{2(g)} + H_2O$$

Typical decomposition results obtained using different solution conditions clearly demonstrate that the BCE process is much more efficient for NH_4HCO_3 decomposition, especially compared with the standard method, of stirring in a water bath. For example, in the BCE process, almost complete thermal decomposition of NH_4HCO_3 can be obtained after 30 minutes of bubbling of 150 °C air through a 0.5 m solution. The initial high concentration of NH_4HCO_3 was found to inhibit bubble coalescence, producing small bubbles. However, after significant reduction in NH_4HCO_3 concentration larger bubbles are formed, as are observed in pure water, which also confirms the decomposition of NH_4HCO_3 salt into ammonia and carbon dioxide gases.

The enhanced thermolysis observed raises some interesting questions about the mechanism of heat transfer which must occur in the BCE process. Preheated gas bubbles, introduced and passed though the aqueous solution, must necessarily produce a transient hot surface layer around each rising bubble. The transient hot surface layer will have a higher temperature than the average temperature of the aqueous solution, and it is believed that it is the interaction of the solute with this transient hot surface layer which results in the thermal decomposition of the solute, even when the average temperature of the aqueous solution remains below the temperature at which it would normally cause thermal/chemical decomposition of the solute. For situations where thermal decomposition is either required very quickly or at reduced temperature, the BCE method offers a new approach.

INHIBITION OF PARTICLE GROWTH IN A BCE

At first, it might appear that the BCE process, with continuous water evaporation via the rising dry bubbles, could be used to slowly increase supersaturation levels for suitable solutes and hence cause precipitation. However, it has been discovered by studying several insoluble aqueous salts (e.g. $CaSO_4$, $SrSO_4$ and $CuCO_3$) that the BCE process actually has a significant inhibition effect on the precipitation process. The initial nanoparticles formed in supersaturated solutions are actually prevented from growing by the presence of the bubbles. By comparison, simple stirring of solutions with the same supersaturation level produces significant particle growth and visually obvious increased turbidity. The mechanisms responsible for this growth inhibition have yet to be identified, but there are clear applications in situations where scale formation presents issues affecting industrial processing.

The bubble column evaporator is clearly a good example of a process which is simple to use and has many potential industrial applications, but it is actually quite difficult to fully understand its precise mechanisms.

SAMPLE QUESTIONS

(1) Estimate the steady state temperature of a 1 M NaCl aqueous solution inside a continuous flow bubble column with dry nitrogen gas flowing into the column through a sinter at an inlet temperature of 165 °C. Assume that the pressure drop across the column is 2000 Pa and that the C_p for N_2 gas is roughly constant at 1.04 J/g K.

(2) Common salt added to a bubble column at the concentration present in human body fluids inhibits bubble coalescence. What do think this effect might have had on human evolution?

(3) In commercial multistage flash (MSF) distillation systems seawater is induced to boil at 80 °C by a reduction in pressure. How will the water vapour collection efficiency compare with a BCE running at the same temperature?

Industrial Report

Bubble column evaporator

The bubble column evaporator is a really simple process to use for many practical operations requiring the transfer of thermal energy and matter. It is robust, easy for low skill operation, low energy consumption and low complexity, but has wide variability of application. Although relatively simple to apply, it is based on a complex level of theoretical understanding of the behaviour of bubbles in water.

Some of the advantages of the process are:

• It maximizes efficiency through maximum gas/solution surface area.

• It forms a natural membrane (e.g. air/water) that can be manipulated or enhanced through different surfactants or carrier gases.

• The temperature of the inlet gas can be raised to increase reaction catalysis or other thermal effects applied to a wide range of aqueous solutes.

• Heated (dry) inlet gas bubbles can be used to sterilize the solution whilst maintaining a moderate temperature.

• Its efficient water vapour capture mechanism can be used for many applications, from seawater desalination to dewatering industrial wastewater slimes.

Amongst the greatest challenges facing many industries today is the efficient, economic removal of low concentrations of specific solutes from water. Extraction of these solutes may be desirable for either economic reasons either because the elements are highly valuable, e.g. rare earth metals, or because the purity of the water itself needs to be improved.

Many thousands of tonnes of otherwise valuable metals are lost in mining operations throughout the world through the disposal of enormous volumes of low concentration metal ions. Whilst careful disposal may alleviate many environmental concerns, the efficient recovery of such material would improve both economic return for miners and costs of environmental management as well as overall environmental risk.

The BCE has a number of advantages that are exploitable in various ways by industry:

- High surface area gas/solution maximises potential rates of reaction or interaction, thereby increasing efficiency
- The gas/solution barrier provides a natural surface interface layer similar to a membrane structure that can be utilised by a range of surfactants. This allows creation of specialized 'membranes' not only with specific properties to target specific applications but potentially with multiple concurrent intervention.
- Natural surface regeneration – continual gas flow allows continual generation of new gas/solution interfaces which prevents clogging and removes the need to regenerate the active surface layer through cleaning or replacement, as would be required for other membrane technologies.

In short, the BCE provides a unique process that has the versatility to be utilised across a wide range of applications with relatively low initial capex costs and low to extremely low operating costs and complexity. The BCE is an example of the power of low-tech solutions which take full advantage of fundamental scientific principles.

Mr Matt Battye
CEO
Breakthrough Technologies
Canberra
Australia

Experiment 11.1   Determination of the enthalpy of vaporization of concentrated salt solutions

Any bubble column evaporator can used to determine the enthalpy of vaporization of concentrated salt solutions. The basic glass apparatus is illustrated below.

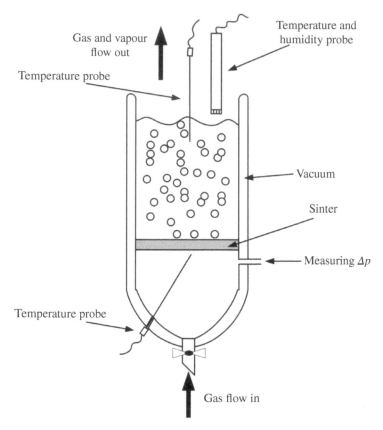

Figure 11.8 Glass apparatus for measuring the enthalpy of vaporization of concentrated aqueous salt solutions.

The key features of the apparatus are a glass sinter (porosity size no. 2) inserted at the base of a glass column, a source of dry gas (e.g. from a N_2 cylinder) and two accurate temperature probes (such as a thermocouple). The flow rate of the gas has to be controlled to give a dense bubble column but doesn't need to be measured. The measured steady-state temperature differential (ΔT) between the inlet gas and the column solution is all that is required experimentally to be able to calculate ΔH_v values using the equation:

$$\left[\Delta T \times C_p\left(T_e\right)\right] + \Delta P = \rho_v\left(T_e\right) \times \Delta H_{vap}\left(T_e\right) \qquad (1)$$

where ΔP can either be measured or estimated by the sum of the hydrostatic pressure of the column and the Laplace pressure created by the pores (of radius r_p)* of the sinter (i.e. $h\rho g + 2\gamma/r_p$). $C_p(T_e)$ is the specific heat of the gas flowing into the bubble column at constant pressure at temperature T_e, which can be obtained from literature values; T_e is the steady-state temperature near the top of the column; ρ_v is the water vapour density at T_e, which can be calculated from literature data on the water vapour pressure of salt solutions at this steady-state temperature, using the ideal gas equation.

For simplicity, a non-vacuum gas bubble column container can be used and simply covered with a suitable insulating material.

A range of concentrated salt solutions can be used, such as seawater (i.e. 0.5 M NaCl) and 3 m $ZnSO_4$ and Li_2SO_4 solutions, which give the widest range of ΔH_v values.

(*Note that the effective r_p for these sinters is about 45 μm.)

Appendices

APPENDIX 1

FUNDAMENTAL CONSTANTS

Avogadro's number, $N_0 = 6.022 \times 10^{23}$ mol^{-1}
Boltzmann constant, $k = 1.381 \times 10^{-23}$ J K^{-1}
Electronic charge, $-e = 1.602 \times 10^{-19}$ C
Permittivity of free space, $\varepsilon_0 = 8.854 \times 10^{-12}$ C^2 J^{-1} m^{-1}
$1kT = 4.12 \times 10^{-21}$ J at 298 K
1atm $= 1.013 \times 10^5$ Nm^{-2} (Pa)
Speed of light $= 2.998 \times 10^8$ m s^{-1}
Planck's constant, $h = 6.626 \times 10^{-34}$ J s
Gas constant, $R = 8.3145$ J K^{-1} mol^{-1}
Rest mass of the proton $= 1.673 \times 10^{-27}$ kg
Molar volume of gas at STP $= 22.414$ L
Average molar mass of dry air $= 29$ g/mol
Gravitational acceleration $= 9.81$ m/s^2

IMPORTANT PROPERTIES OF WATER

Viscosity of water $= 0.001$ N s m^{-2} at 20 °C
Viscosity of water $= 0.00089$ N s m^{-2} at 25 °C

Applied Colloid and Surface Chemistry, Second Edition. Richard M. Pashley
and Marilyn E. Karaman.
© 2021 John Wiley & Sons Ltd. Published 2021 by John Wiley & Sons Ltd.
Companion website: www.wiley.com/go/pashley/appliedcolloid2e

Dielectric constant of water = 80.2 at 20 °C
Dielectric constant of water = 78.5 at 25 °C

0 °C = 273.15 K, triple point for water = 273.16 K

pH of the ocean = 8.1 (was 8.2 prior to industrialization)
pH of human blood = 7.4 (+/− 0.1).
pH of rainwater = 5.5–5.6 (due to dissolved CO_2)
pH of tap water = 6.5–8.5

USEFUL INFORMATION IN SURFACE CHEMISTRY

kT/e = 25.7 mV at 298 K
1 C m^{-2} = 1 unit charge per 0.16 nm^2

ZETA POTENTIALS

From microelectrophoresis measurements on spherical colloidal particles, dispersed in water, the observed average electromobility U_E (in units of μm s^{-1}/V cm^{-1}) is directly related to the particle's zeta potential by the equation:

$$\zeta (mV) = 12.8 \times U_E \qquad \text{at } 25°C$$

DEBYE LENGTHS

$$\kappa^{-1}(nm) = \frac{1.0586 \times 10^{13}}{\left[\Sigma_i C_i(B) Z_i^2 \right]^{1/2}} \qquad \text{at } 21°C, \text{with } C_i(B) \text{ values in nos/m}^3$$

$$\kappa^{-1}(nm) = \frac{0.305}{\sqrt{M}} \qquad \text{for 1:1 electrolytes, where M is molarity at } 21°C$$

For example:

Conc NaCl/M	0.1	0.01	0.001	0.0001
κ−^{1}/nm	0.96	3.05	9.64	30.5

and

$$\kappa^{-1}(nm) = \frac{0.176}{\sqrt{M}} \qquad \text{for 2:1 electrolytes} \left(\text{at } 21°C \right)$$

SURFACE CHARGE DENSITY

The corresponding surface charge density on a flat, isolated surface of potential ψ_0 immersed in 1:1 aqueous electrolyte solution, at 25°C, is given by the relation:

$$\sigma_0\left(Cm^{-2}\right) = \frac{3.571 \times 10^{-11}}{\left(\kappa^{-1}/m\right)} * sinh\left(\psi_0/mV \times 0.01946\right)$$

APPENDIX 2

MATHEMATICAL NOTES ON THE POISSON-BOLTZMANN EQUATION.

The solution to Equation (8.12) in Chapter 8: $\dfrac{d^2Y}{dX^2} = sinh\,Y$ depends upon the boundary conditions of the system under consideration. The two main cases of interest are (a) an isolated surface and (b) interacting surfaces. In both cases we can carry out the first integral of Equation (8.12) using the identity:

$$\int \frac{d^2y}{dx^2}\,dy = 0.5\left(\frac{dy}{dx}\right)^2 + C$$

and hence:

$$\left(\frac{dY}{dX}\right)^2 = 2\cosh Y + C \tag{1}$$

Determination of the integration constant C depends on the boundary conditions. For case (a) an isolated surface:

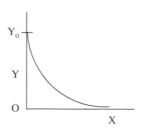

$dY/dX = 0$ when $Y = 0$ and hence $C = -2$ and equation (1) becomes:

$$\left(\frac{dY}{dX}\right)^2 = 2\left(\cosh Y - 1\right) \tag{2}$$

Equation (2) can be integrated to obtain an exact analytical equation for the potential decay away from a flat surface

The simplest example of case (b) is for the interaction of identical surfaces i.e. the symmetrical case shown below:

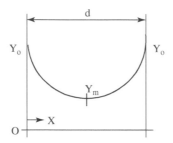

In this case, we can use the boundary condition $(dY/dX) = 0$ when $Y = Y_m$ (the mid-plane potential) and hence $C = -2\cosh Y_m$. Equation (1) then becomes:

$$\left(\frac{dY}{dX}\right)^2 = 2\left(\cosh Y - \cosh Y_m\right) \tag{3}$$

Unfortunately, integration of this equation is rather difficult and leads to elliptic integrals which only have numerical solutions. (A relatively simple numerical solution to Equation (3), without the use of elliptic integrals, was developed by Chan et al., *J. Colloid Interface Science* 77(1): 283, 1980.)

Even though this equation is difficult to solve, many approximate methods have been used. Equation (3) is, however, interesting for what it tells us about the double-layer interaction. It can be rearranged in the form:

$$\left(\frac{dY}{dX}\right)^2 - 2\cosh Y = 2\cosh Y_m \neq f\left(Y \text{ or } X\right) \tag{4}$$

Clearly, the sum of the two terms on the left-hand side is constant at any point between the two interacting surfaces.

We can now identify the first term in equation (4) with Maxwell's stress tensor, which acts on any dielectric in an electric field. The magnitude of this force $|F_E|$ is given by:

$$|F_E| = \frac{\varepsilon_o D|E|^2}{2} \tag{5}$$

where the electric field strength $|E|$ is proportional to dY/dX. This electrostatic force acts on each surface to pull it towards its oppositely charged diffuse layer. The second force is due to the osmotic pressure generated by the excess of (charged) solute counterions in the interlayer between the interacting surfaces, compared with the bulk solution. This pressure acts to push the surfaces apart and is proportional to the (total) local electrolyte concentration. It can be easily demonstrated that the local electrolyte concentration is in turn proportional to $cosh Y$, which is actually the sum of the Boltzmann factors for each ion.

It follows that for interacting flat surfaces at any given separation distance, two opposing forces – electrostatic and osmotic – vary in magnitude across the liquid film but compensate each other to give the overall repulsive pressure.

For the symmetrical case, it is easier to obtain the value of this pressure using the fact that at the mid-plane, where $Y = Y_m$, the electrostatic term disappears (i.e. $dY/dX = 0$, $|E| = 0$), and the total pressure is equal to just the osmotic pressure at this plane, which can then be easily calculated once the value of Y_m is known.

APPENDIX 3

NOTES ON 3-D DIFFERENTIAL CALCULUS AND THE FUNDAMENTAL EQUATIONS OF ELECTROSTATICS.

Div: $\quad \vec{\nabla} \bullet \vec{E}(r) = \dfrac{\rho(\vec{r})}{\varepsilon_0 D} = \dfrac{\partial E_x}{\partial x} + \dfrac{\partial E_y}{\partial y} + \dfrac{\partial E_z}{\partial z} \quad$ (a scalar)

Curl: $\quad \vec{\nabla} \times \vec{E}(\vec{r}) = 0$

Grad: $\quad \vec{E}(\vec{r}) = -\vec{\nabla}\psi(\vec{r}) \equiv -\left(\dfrac{\partial \psi}{\partial x}, \dfrac{\partial \psi}{\partial y}, \dfrac{\partial \psi}{\partial z}\right), \quad$ i.e. the vector i_x, i_y, i_z

$$\vec{E}(\vec{r}) = \lim_{q=0} \frac{\vec{F}(\vec{r})}{q}$$

GAUSS'S LAW:

$$\oint_{surface} \vec{E} \bullet \hat{n} da = \frac{\overset{insurface}{\underset{i}{\sum}} q_i}{\varepsilon_0}$$

Index

Page locators in **bold** indicate tables. Page locators in *italics* indicate figures. This index uses letter-by-letter alphabetization.

Applied Colloid and Surface Chemistry, Second Edition. Richard M. Pashley
and Marilyn E. Karaman.
© 2021 John Wiley & Sons Ltd. Published 2021 by John Wiley & Sons Ltd.
Companion website: www.wiley.com/go/pashley/appliedcolloid2e